Topics in Information Systems

Editors:

Michael L. Brodie
John Mylopoulos
Joachim W. Schmidt

Springer Series
Topics in Information Systems

M. L. Brodie, J. Mylopoulos, J. W. Schmidt (Eds.):
On Conceptual Modelling: Perspectives from Artificial
Intelligence, Databases and Programming Languages.
XI, 510 pages, 25 figs., 1984

W. Kim, D. S. Reiner, D. S. Batory (Eds.): Query Processing in
Database Systems. XIV, 365 pages, 127 figs., 1985

D. C. Tsichritzis (Ed.): Office Automation: Concepts and Tools.
XII, 441 pages, 86 figs., 1985

M. L. Brodie, J. Mylopoulos (Eds.): On Knowledge Base
Management Systems: Integrating Artificial Intelligence and
Database Technologies. XXI, 660 pages, 89 figs., 1986

L. Bolc, M. Jarke (Eds.): Cooperative Interfaces to Information
Systems. XIV, 328 pages, 62 figs., 1986

M. P. Atkinson, P. Buneman, R. Morrison (Eds.): Data Types
and Persistence. XVI, 292 pages, 10 figs., 1988

Data Types and Persistence

Edited by

Malcolm P. Atkinson
Peter Buneman
Ronald Morrison

With 10 Figures

Springer-Verlag
Berlin Heidelberg New York
London Paris Tokyo

Series Editors

Michael L. Brodie
GTE Laboratories Incorporated
40 Sylvan Road, Waltham, MA 02254, USA

John Mylopoulos
Department of Computer Science, University of Toronto
Toronto, Ontario M5S 1A7, Canada

Joachim W. Schmidt
Fachbereich Informatik, Johann Wolfgang Goethe-Universität
Robert-Mayer-Straße 11–15,
D-6000 Frankfurt a. M. 11, Fed. Rep. of Germany

Volume Editors

Malcolm P. Atkinson
Department of Computing Science, University of Glasgow
14, Lilybank Gardens, Glasgow G12 8QQ, Scotland

Peter Buneman
Department of Computer Information Science, The Moore School
of Electrical Engineering, University of Pennsylvania
Philadelphia, PA 19104-3897, USA

Ronald Morrison
Department of Computational Science, University of St. Andrews
North Haugh, St. Andrews, Fife KY16 9SS, Scotland

ISBN 3-540-18785-5 Springer-Verlag Berlin Heidelberg New York
ISBN 0-387-18785-5 Springer-Verlag New York Berlin Heidelberg

Library of Congress Cataloging-in-Publication Data.
Data types and persistence. (Topics in information systems) "Proceedings of a workshop held at the Appin in August 1985"–Pref. Bibliography: p. Includes index.
1. Data structures (Computer science)–Congresses.   2. Data base management–Congresses.   3. Programming languages (Electronic computers)–Congresses.   I. Atkinson, M. P.   II. Buneman, Peter, 1943-   .   III. Morrison, R. (Ronald), 1946-
IV. Series.   QA76.9.D35D39   1988   005.7'3   88-4606
ISBN 0-387-18785-5 (U.S.)

Printing: Druckhaus Beltz, Hemsbach/Bergstr.;
binding: J. Schäffer GmbH & Co. KG, Grünstadt
2145/3140-543210

Topics in Information Systems

Series Editors:

Michael L. Brodie, Computer and Intelligent Systems Laboratory, GTE Labs.
John Mylopoulos, Department of Computer Science, University of Toronto
Joachim W. Schmidt, Department of Computer Science, University of Frankfurt

Series Description

Computer Science is increasingly challenged to deliver powerful concepts, techniques, and tools for building high quality, low cost information systems. In the future, such systems will be expected to acquire, maintain, retrieve, manipulate, and present many different kinds of information. Requirements such as user-friendly interfaces, powerful reasoning capabilities, shared access to large information bases, and cooperative problem solving all with high performance are becoming ever more popular to potential information system users. More fundamentally, there is an ever increasing need for powerful environments for the design and development of such information systems.

Software technology for building these systems is far from meeting these requirements. Despite major achievements in every area of Computer Science, the gap between what is expected and what Information System technology can deliver is widening. This is in marked contrast to dramatic advances in individual research areas such as hardware technology, knowledge-based systems, distributed processing, graphics, user-interfaces, etc. The critical challenge in meeting the demand for high quality, low cost information systems cannot be addressed successfully by individual technologies. Rather, it critically depends on our ability to integrate technologies. One reason for the urgency of this task is that dramatic advances in hardware technology have made computers available to an ever-growing community of potential users all of whom have definite information processing needs.

The Topics in Information Systems (TIS) series of books focuses on the critical challenges of technology integration for information systems. Volumes in the series will report recent significant contributions to the conceptual foundation, the architectural design, and the software realization of information systems. The series is based on the premise that these tasks can be solved only by integrating currently distinct technologies from different areas of Computer Science such as Artificial Intelligence, Databases and Programming Languages. The required dramatic improvements in software productivity will come from advanced application development environments based on powerful new techniques and languages. The resulting new technologies should allow us to transform our conceptions of an application domain more directly and efficiently into operational systems utilizing conceptual modelling methodologies, new languages for requirements, design and implementation, novel environments, performance analysis tools, and optimization techniques.

The *concepts, techniques,* and *tools* necessary for the design, implementation, and use of future information systems is expected to result from the integration of those being developed and used in currently disjoint areas of Computer Science. Several areas bring their unique viewpoint and technologies to existing information processing practice. One key area is *Artificial Intelligence* (AI) which provides knowledge representation and reasoning capabilities for knowledge bases grounded on semantic theories of information for correct interpretation. An equally important area is *Databases* which provides means for building and maintaining large, shared distributed databases grounded in computational theories of information for efficient processing. A third important area is *Programming Languages* which provides a powerful tool kit for the construction of large, efficient programs and programming environments to support software engineering. To meet evolving information systems requirements, additional research viewpoints and technologies are or will be required from such areas as *Software Engineering, Computer Networks, Machine Architectures* and *Office Automation.*

Although some technological integration has already been achieved, a quantum leap is needed to meet the demand for future information systems. This integration is one of the major challenges for Computer Science in the 1990s.

The TIS series logo symbolizes the scope of topics to be covered and the basic theme of integration. The logo will appear on each book to indicate the topics addressed. A recent book in the series, *On Knowledge Base Management Systems: Integrating Artificial Intelligence and Database Technologies,* edited by Michael L. Brodie and John Mylopoulos, for example, deals with concepts and techniques in AI and Databases and has the logo

	Artificial Intelligence	Databases	Programming Languages
Concepts	●	●	
Techniques	●	●	
Tools			

All books in the series provide timely accounts of ongoing research efforts to reshape technologies intended for information system development.

Michael L. Brodie
John Mylopoulos
Joachim W. Schmidt

Preface

There is an established interest in integrating databases and programming languages. This book on Data Types and Persistence evolved from the proceedings of a workshop held at the Appin in August 1985. The purpose of the Appin workshop was to focus on these two aspects: persistence and data types, and to bring together people from various disciplines who have thought about these problems. Particular topics of interest include the design of type systems appropriate for database work, the representation of persistent objects such as data types and modules, and the provision of orthogonal persistence and certain aspects of transactions and concurrency.

The programme was broken into three sessions: morning, late afternoon and evening to allow the participants to take advantage of two beautiful days in the Scottish Highlands.

The financial assistance of the Science and Engineering Research Council, the National Science Foundation and International Computers Ltd. is gratefully acknowledged. We would also like to thank Isabel Graham, Anne Donnelly and Estelle Taylor for their help in organising the workshop. Finally our thanks to Pete Bailey, Ray Carick and Dave Munro for the immense task they undertook in typesetting the book.

The convergence of programming languages and databases to a coherent and consistent whole requires ideas from, and adjustment in, both intellectual camps. The first group of chapters in this book present ideas and adjustments coming from the programming language research community. This community frequently discusses types and uses them as a framework for other discussions.

Types in programming languages were first introduced to describe the values that a program might manipulate, for example INTEGER and REAL in Fortran. They could then be used to specify which values could be manipulated by which operations. Initially a fixed set of types was provided. Later the programmer was provided with the facilities to define types using a fixed repertoire of type constructors (e.g. **set, array, record**) which could be parameterised with other types to yield new types. Pascal [JW78] and Algol 68 [VANW75] are examples of this genre.

This introduces a major transition in the use of types. Initially the language designer defines types to name value sets and then describes their use. Later the *programmer* also defines types, using a small number of recipes defined by the language designer. This enables the programmer to better describe the data structures to be manipulated. A number of rôles are now played by types. The pre-definition of data structures allows application relevant names of parts of a data structure to be introduced. The more explicit type specification implies more precise limits on the matching of functions (or operations) and their operands. This increases the number of programming errors that can be detected by the compiler, such errors are known to cause problems in much database application programming. Finally better description of the intended data structure provides information to the compiler and run time system so that they can (in principle) support the structures more efficiently.

In the above type systems the language designer still dictates the kinds of type that the programmer may define. Two developments have reduced this restriction. The introduction of abstract data types allows the programmer to separately define the behaviour and (possibly multiple) implementations of those types. One motive for this, particularly when programming in the large, is the hiding of the implementation to allow independent development within the constraints of an agreed interface. The other motive is to allow the programmer to define a new component for constructing data structures - this component may then be used by other programmers as well. The second development introduced to overcome an irksome restriction in the type systems of the previous paragraph is parametric polymorphism [MILN78] [CW85]. In the preceding type system, it was possible to construct different types, such as **list of** integer, **list of** person, and then when identical operations were required, such as *reverseList*, *sortList*, *lengthOfList*, the programmer had to write out the code for each type of operand. Parametric polymorphism allows types such as **list of a** (where a is any type to be provided later) to be defined, and operators over them can be defined which work for any consistent substitution for a. ML, Galileo [Chapter 8] and Poly [Chapter 4], are good examples of such languages, where the combination of polymorphism and abstract data types allows the programmer to define type constructors.

Although the first section of this book is explicitly concerned with types, they are an issue in many other chapters of the book. As we are concerned with database programming languages the issue of types can arise in a number of ways.

First they can arise as the language designer explains the values that may be manipulated, particularly those that are introduced to model the regularity of large scale structures, such as relations and databases in Pascal/R [SCHM77] and its descendents [KMPRSZ83] [EEKMS85]. The major motive there is to provide a predefined data structure building component which it would be difficult for a programmer to define. The information conveyed to the run time system allows significant optimisation of the operations on data held on disc. This is achieved at the cost of staying with the Pascal/Algol-68 genre of type systems and being quite restrictive about the type parameterisations allowed. The proposals for Napier [Chapter 1] make an attempt to combine the index and relation types with complete freedom of type parameterisation and polymorphism.

The design of database programming languages (DBPLs) must acknowledge that very large application software suites will be constructed to operate against the data. This is met by introducing abstract data types (ADTs), and modules. ADTs are part of the necessary type structure in Napier [Chapter 1], and appear similarly in Poly [Chapter 4]. Cardelli and McQueen examine three interpretations of the

ADT concept [Chapter 3]. Clarity about such issues, which for example determine whether the instances (and types) of two ADTs may be tested for equality, is essential in language design though it may be taken for granted by the consumers (programmers).

Related to ADTs are modules which are the components from which a large system is built. They are naturally stored in the persistent store of a DBPL. As procedures, this is already the practice in PS-algol [AM85a] and Napier would continue this tradition [Chapter 1], storing both procedures and ADTs in persistent data structures. ML requires the introduction of a new construct to organise such modules [Chapter 2] with functors as a means of organising their composition and hence program construction.

As a DBPL is used against long lived data, then some of the software will be required to operate for many years. This raises the need to support both software and data evolution, which requires new types of binding. Napier's designers believe that the ubiquitous provision of mechanisms for evolution would prove too inefficient, and so provide the programmer with the ability to specify when binding should be performed [Chapter 1]. Thus points of flexibility are introduced. The issue behind this is the choice between dynamic and static binding and consequently dynamic and static typechecking. The designers of many languages have sought complete static typechecking (all the type rules may be checked by inspection of the program source), they argue this is safer and more efficient. Other designers have chosen dynamic binding and type checking everywhere, for example Smalltalk [GR83] and Lisp [MCAR62].

For persistence it is generally agreed that a judicious mixture of static and dynamic binding is required. This was provided by type **pntr** in PS-algol, and it is replaced by type namespace in Napier [Chapter 1]. In Amber [CARD85] it is referred to as type **dynamic** and the operations on this type are described by Cardelli & MacQueen [Chapter 3].

A technical problem in the search for a good type system is the implementation of type checking. As the type language permits increasing refinement of the specification of types, and the type rules permit controlled polymorphism, so the task of typechecking becomes more challenging. Fairbairn [Chapter 6] describes such a type checker, for a functional programming language Ponder. The type systems are all viewed as imposing a structure (naming subsets) on a universe of values. This universe is infinite, and through parameterisation, recursion and polymorphism, so are the subsets identified by many types. Two types are equivalent if they identify the same subset, and often the function application rules require the type checker to establish either that the provided type (for an actual parameter) is equivalent to the specified type (for the formal parameter) or that it is a subset. To make this computationally feasible requires certain constraints on the type system.

One such constraint is to separate the universe of types from the universe of values. The language Poly [Chapter 4] allows some manipulation of types, for example they may be produced by procedures, and hence be data dependent, but care is taken to avoid operators that would violate this separation. In Poly the definition of polymorphic constructs includes a means of specifying constraints on the type parameters. A signature may be specified for each type parameter to ensure that the substituted type provides certain operations. A similar use of signatures in Napier [Chapter 1] is used to ensure that a dynamically bound name space provides the expected names and values.

A fundamental interest in type systems for DBPLs is their tantalising similarity to data models. Data models are type constructors, which when parameterised via some data description language (a schema), which is much like a type language, yield a type which describes the set of all possible databases (values) compliant with that schema. The motives for a type system and a data model are the same: to describe and name the data for benefit of future discourse, to allow rules implementing partial correctness to be specified and automatically checked, and to convey information to the system so it may better implement the required data structures. We therefore expect the two domains to inform one another, and, in the longer term, might expect a data model to be a defineable abstract data type. These ideas underly many of the chapters in this book, particularly, relations from databases have their influence in Napier [Chapter 1], inheritance and classes from databases influences Galileo [Chapter 8] and an attempt to provide a theoretical framework for introducing relations consistent with records is presented by Buneman [Chapter 7]. Similiarly the Semantic Data Models and Functional Data Models have influenced Galileo [Chapter 8].

To achieve a consistent treatment of type in long term and short term data is clearly advantageous, as it removes a barrier that the programmer currently has to overcome. But there are practical difficulties, discussed in more detail in the later sections of the book. The question that faces the type theorist and language designer is "is the same information required and adequate for organising the support of short lived and long lived data structures?" Those that only permit relations to persist [Chapter 12] [SCHM78] [SFL81] believe not, whereas the implementors of Napier [Chapter 1], Galileo [Chapter 16] and Poly [Chapter 4] believe it may be sufficient.

Different arrangements may be made to achieve persistence. That for Napier is consistent with the work on PS-algol, and is presented in later sections of the book. Poly uses a similar incremental loading policy, and commits at the end of the session. Three strategies are reviewed by Cardelli and MacQueen [Chapter 3]. There is an interplay between the type system and the persistence. An example is the requirements for delayed binding already discussed. Another example is the treatment of sharing.

'Sharing' is used in two ways in the literature. Many programming language designers use it to describe references to the same substructure from more than one place in a data structure. This has an impact on the semantics if the substructure is mutable. Some of the persistence mechanisms [Chapter 3] do not preserve such sharing. In database parlance 'sharing' refers to the concurrent use of mutable data structures by different processes. If the data structure representing the entire computation is considered, then each process will be a data structure which during the time of use must contain a reference to the common substructure. Hence the two ideas are related.

In functional programming languages the data structures are not mutable and hence both forms of sharing are not available. This precludes the use of functional languages as DBPLs. Nikhil [Chapter 5] explores ways of reintroducing mutability and sharing into a functional language in a controlled way so that it may serve as a DBPL. He has shown that higher order functions (also available in ML, Napier, Poly etc.) provide an expressive notation for describing database queries. In the strictly applicative language much program transformation and hence optimisation is possible, which may give significant performance improvement for disc data structures. The challenge is to introduce mutability and sharing without compromising this potential for transformation.

Throughout the first section of this book there is a search for a good compromise type system for DBPLs. There are trade offs between implementation problems and consistent type notations; between expressiveness and type checking and between responding to programming language and to database practice. Just as the type system in a DBPL is the analogue of data model in data bases, the engineering decision of what to statically determine about the program or database, and what to check dynamically remains an issue in both domains of research. Also in both cases it is as yet unresolved whether the specification of invariants or constraints (those aspect of correctness we can formalise and may be able to check) requires a separate notation.

As we noted in the introduction, the major reason for a discussion of types and persistence is to obtain a better understanding of database programming languages. The first section of the book was concerned with general problems of types and persistence, this section is mainly concerned with how data bases actually get represented as types in programming languages. The ideas presented here go beyond the survey [ATKI85], and should not necessarily be thought of as detailed language specifications (though in many cases they are just that) but more importantly as ideas about type systems that could contribute to the design of future database programming languages.

The history of database programming languages extends over twelve years. The earliest, and perhaps still the most successful, is Pascal-R [SCHM77]. At the same time, a number of other proposals and implementations emerged [ROWE79] [WASS81]. Given the obvious utility of such languages, one wonders why the latter category did not obtain more widespread acceptance. A possible reason is that the type system — the motivation for the design of these languages — actually got in the way of practical programs. For example, as argued in chapter 1 we must dispense with our conventional notion of static type checking. A general-purpose database query language or application (such as a "forms" system) cannot be implemented in a language in which the types are completely specified in advance. To a certain degree these problems can be overcome using the kinds of type parameterization discussed in chapter 6 and chapter 4. However, in the case of, say a forms generating system there are two stages at which one would like to see the type checking take place: first when the program is compiled, and second at the time a forms system is generated. Note that we would hope that the end user would not be able to enter data that could create type errors - from this point of view the checking is static. However it is not static in the conventional sense; "incremental" might be a better term.

Another drawback to these early approaches was that the type system failed to take account of inheritance. The idea had been around for some time in the semantic networks of Artificial Intelligence, but its relevance to databases was first recognized in [SMIT77] and [HAMM81]. This led to a number of attempts notably Adaplex [SMIT81], Taxis [MYLO80] and Galileo [ALBA83], all of which are being implemented, to incorporate some form of inheritance into their types. In doing this, they come close to object-oriented languages in associating classes, or some form of extent, with types. Galileo is the subject of chapter 8 by Albano and his colleagues. Based on ML [GORD79] it has the most precisely defined notion of type, and in this chapter a detailed account of the type constructors and checking is given. Of special interest are the use of polymorphism and the support for type abstraction.

If we have a mechanism for maintaining persistent objects, then it may be possible to use such objects as an implementation for modules. This is suggested in [ATK85] and explained in more detail in chapter 11. The idea is that a program itself may be represented as a structured object, and one can store this and use it as a module if one has a suitable underlying database management system. Zdonik and Wegner use the term "databased programming languages" to refer to this method of language construction. They give other examples of how a persistent object-oriented approach can deal with other structures such as file systems.

Despite the flexibility of an object-oriented approach or, more generally, an approach based upon type inheritance, there is always the need to accommodate exceptional data. Typically this happens in databases when attempts are made to represent a greater variety of data than that for which the database is designed, but it is a problem in any environment where the design task is complicated or the designer has incomplete information about the structures to be represented. Borgida points out that there should be some similarity to exceptions in databases and those in programs, the distinction being that the former persist. In chapter 10 he shows that these exceptions typically violate type constraints and suggests mechanisms for accommodating such violations without their effects becoming too widespread, i.e. without having to modify large numbers of applications that run against the database and assume that the data conform to certain types. To implement this the addition of some dynamic type-modifying mechanisms are certainly required and need further study.

A well-known definition of a data type is that it is a *signature*, roughly speaking a set of operations with the appropriate arity. If this is the case then it would be appropriate to study the operations that are well-known in databases and this might lead us to an appropriate description of the type. A prime example of such a set of operations is the relational algebra; and in chapter 12 it shows that if one extends the relational algebra with a few additional operations, one effectively has a full programming language. Moreover this language is appropriate for more than database work. Just as APL, originally designed for scientific tasks involving array processing [FALK83], was useful for many other programming tasks; so also can Aldat be used for non-relational tasks as diverse as computational geometry and text-processing. Whether such an approach can easily be used for other database applications, such as the generation of forms mentioned above remains to be seen.

The two approaches to types: relational algebra and inheritance appear, at first site to be completely opposed. Languages such as Galileo may have some ability to manipulate extents, but these do not usually exceed the bounds of what is possible with sets, and do not approach the richness of type relational algebra. In an attempt to add typed extensional functions (to represent indexes and other structures that are essential in database programming) Buneman, in chapter 7, shows that at least some relational operators such as natural join emerge quite naturally from operations on functions. As a result one can easily assign a data type to natural join using a rule for typing functions on ordered types originally worked out by Cardelli in [CARD84]. It is therefore possible that relational algebra and object-oriented programming may not be opposite approaches, but may both be representable in some more general framework.

Another area of programmming language which has, until recently, denied any notion of inheritance is logic programming. Although the idea that a relation can be represented by a set of predicates [GALL78] is extremely powerful and appealing, it is difficult to see how, at a syntactic level, one can apply this idea when one has relations with, say, thirty or more columns. This is not uncommon in the real-world of database applications. Using an orderings on labelled structures, which has a close relationship with the orderings used on types, Ait-Kaci chapter 9 shows how the notion of inheritance works smoothly in the domain of logic programming. Better, the interpreter for his language has an algebraic representation that is closely related to the algebra used in relational databases.

If, therefore, there is a unifying point to these papers that deal with the representation of databases in programming languages, it is that inheritance — rather than being an added complication that we need to impose on our type systems — is something that offers to unify a number of disparate ideas. Of course, what is presented here is only a start. Among other things we need to improve greatly our understanding of the semantic issues involved.

One aspect of large bodies is that often there is a requirement to access them concurrently. The requirement arises from the need to operate the database efficiently and to allow the programmer to model using concurrent processes. Concurrent access to data raises difficulties in understanding the semantics of concurrency, distribution, sharing, transactions and stability of the data. Models of concurrency, [MILN80] and language constructs to support concurrency, such as semaphones [DIJK68]], conditional critical regions [BRIN75], monitors [HOAR74], path expressions [CH74], rendezvous [ICHB83] etc., have been proposed but have always stopped short of a total solution. This highlights the complexity of the problem and perhaps should serve as a warning to us that there may not be a simple model or solution to concurrency. In the first part of this section we concentrate on the need to protect and provide concurrent access to the data.

The two papers that are concerned with concurrency both concentrate on transactions. The need for transactions was first identified in work on databases [DAVI73] where it was seen as advantageous that the total database should maintain its consistency by making the changes caused by a transaction atomic; that is the transaction either takes place (commits) or it does not (aborts): there is no middle state. Atomic transactions are characterised by two properties - serialisation and recovery. Serialisation specifies that within a group of concurrent processes acting on shared data, the result is as if the processes had executed serially in some order.

Recovery ensures that if the transaction fails to commit then the database returns to some consistent state. Both serialisation and recovery have to take account of shared data in concurrent transactions. In particular adequate locking protocols are required to preserve determinacy. These locking protocols will also be used in recovery when locks are released and in the prevention of deadlock.

In most systems, transactions are committed on a stable store such as disk and thus serve two purposes by providing both a stability mechanism as well as a method of viewing consistent forms of the data. It is, however, not necessary to couple these two concepts.

Nested transactions or sub-transactions are used to decompose large transactions into smaller units. The concept is clearly composable from the computer's atomic store update. In his paper Weihl [Chapter 13] uses nested transactions and the example of a semi-queue to highlight two approaches to building transaction based systems. In the first method, that of Argus [LISK83] he uses the implicit approach to commitment where the system is responsible for updating the representation of the data. By constructing atomic abstract data types using firstly serial specifications and then behavioural specifications to control the concurrency, he shows how concurrently accessed semi-queue might be built. The locking in the system is provided by a built-in system primitive but it is the responsibility of the user to implement the locking correctly.

Weihl demonstrates that not having explicit control over commit and abort can lead to difficulties in expressing some kinds of constraints. He goes on to propose an explicit approach to concurrency, provides language constructs, and then constructs the semi-queue example. The result is still complex and much evaluation of the solution is required but Weihl deems the systems to have some advantages.

Krablin also investigates an explicit approach to commitment. He describes extensions to PS-algol [PS85] for the concurrent language CPS-algol [KRAB85] which he uses to demonstrate the implementation of two phase commit protocols and sub or nested transactions. Krablin uses Weihl's semi-queue example and it is interesting to compare the two solutions.

In the second part of this section we turn our attention to machine design and in particular to the impact of orthogonal persistence on computer architecture. Cockshott starts off the section by reviewing the addressing strategies that have been used and have been found successful in the implementation of persistent stores. These range through textual flattening, core dumping, paged vertical memory, segmented virtual memory, associative PID addressing with paged virtual memory and finally multiple address space models. Cockshott uses his survey to justify a proposal for associative object addressing with a paged virtual memory.

In the second paper, which was given by Georgio Ghelli, the implementation of persistence for the language Galileo [Chapter 16] is described. Recovery and reliability in the system is provided at 3 levels. An undo log allows recovery from user failures, a shadow page mechanism for failures of software and incremental dumping for hardware failures. The address mechanism used is, in Cockshott's classification, a paged virtual memory with multiple address spaces. The nature of garbage collection in persistent systems constitutes a major part of this paper. An algorithm based on Hewitt & Leiberman [HL83] which divides the store into sections according to their expected lifetime is used. This is particularly appropriate for persistent system since they abstract over this very aspect of data. An improvement to the algorithm based on reference counts but which will collect cycles [Chapter 16] is also described.

Quite clearly there is a lot of research, experimentation and measurement to be performed before we will have clear implementation strategies for persistent systems. Methods of implementation for languages like Galileo or PS-algol have been constructed, the next step is for efficient, perhaps self adjusting implementations.

Judging from present progress, there remain a number of intractable problems in concurrency. Much work is being done, in both the programming language and database communities, in trying to understand the nature of concurrency and provide a conceptual model for it. We see clearly that in certain cases a particular model works well but unifying model has escaped us. The nagging doubt is that there is no guarantee that a simple solution exists.

Table of Contents

Part I

Data Types and Persistence

Chapter 1

Types, Bindings and Parameters in a Persistent Environment

Malcolm P. Atkinson
University of Glasgow

Ronald Morrison
University of St Andrews

1. INTRODUCTION

Our experience of persistent data in PS-algol [ABCCM83] and surveys we have conducted [AB85] [ABCCM81] have led us to identify various requirements on the stores embodied in the design of a programming language. Axiomatic in our approach to language design are:

(i)   languages should be strictly typed;

(ii)   type systems can be discovered that will permit total systems to be built, strictly typed throughout;

(iii)   languages should eventually provide convenient and consistent notations for **all** activities inherent in programming.

This leads to a search for constructs that handle activities currently not supported. The requirements identified for persistent data are:

(i)   the rights to transience or longevity are completely independent of the type of data;

(ii)   the data types should be expressive enough to capture the structure and regularity that will be required of the data;

(iii)   the data types should include program, as procedures and abstract data types;

(iv)   the data types should include a general purpose indexing mechanism, both polymorphic and variadic;

(v)   the data types should include a bulk data type with operations on collections of data;

(vi)   the data types should include some form of inheritance to model specialisation;

(vii)   the types and bindings used should allow system evolution on an incremental and localised basis;

(viii) there should be mechanisms to permit and control shared and concurrent usage;

(ix)   it should be possible to encapsulate any sequence of operations on the store into a transaction, which behaves as a single operation;

(x)   there should be privacy and access control mechanisms.

Currently we are engaged in the design of a new language, Napier, to succeed PS-algol. This paper presents some aspects of the design of the binding and type checking we are considering. These particularly address requirements vi and vii above. Requirement vii is equivalent to the requirement for data independence:schema editing without having to recode, recompile or relink all application programs, a requirement identified by the database community in the early 1970s. It is addressed in our design by controlled and localised incremental binding and type checking, in the context of a structure we call name spaces. In compliance with requirement i, we do not constrain its use to long term data, which implies that it may also be used to allow parts of a program to evolve, and may be used to construct flexible interfaces within a program.

We are able to show that these name spaces, structures, records, abstract data types and relations may all be put into a consistent framework described in terms of associating a formal store with an actual store, similar to parameter mechanisms. We also show that this enables, at least in principle, many activities previously dealt with outside the programming language, to be supported from within it.

The requirement of data description, is met by a polymorphic type system, discriminated unions and abstract data types. This is described in detail elsewhere, but shown to be consistent with name spaces in sections 5 and 6 of the paper. In particular, the type matching based on signatures, and the provision of an environment by an ADT are made to have syntactic and semantic similarities with the matching and binding of name spaces. The projection out of a discriminated union to a specific type is shown to have a similar dynamic binding and checking requirement to the dynamic use of name spaces.

The benefits and mechanisms for providing program as part of the persistent type system we have reported elsewhere [AM84].

A relational type constructor, without restrictions on its component types, provides for indexing, bulk data structures, operations and functions stored by enumerating their results. This is briefly described in section 7.

2. BINDING

In programming language systems there are a number of schemes for binding names of objects. Landin [LAND66] proposed the Principle of Correspondence where the methods of binding in any language are the same for both declarations and parameters. More formally Tennent [TENN77] stated that for any formal parameter F (ie the name being declared) and compatible actual parameter A (the value), the effect on the block body of the qualifying declaration F=A should be identical to an abstraction body having F in its formal parameter list when A is the corresponding actual parameter. For elegant languages there should be a one-to-one correspondence between the methods of binding names to objects in declarations and parameters.

This one-to-one correspondence includes methods of parameter passing (call by value, call by reference etc.) and time of binding (static or dynamic). For the present we will concentrate on time of binding.

In static binding the association between name and object can be determined by a static analysis of the program. The duplicate but legal use of names can be resolved by the static scope rules. For example

```
let  a := 3 ; let  b := 4          ! a contains 3, b contains 4
begin
        let  a := 16               ! a contains 16, b contains 4
        . . .
        b := a * b                 ! a contains 16, b contains 64
        begin
              let b := 93          ! a contains 16, b contains 93
              . . .
              a := a + b           ! a contains 109, b contains 93
        end
end
```

This type of binding was first introduced by algol 60 [NAUR63] and has been continued in the algol tradition by languages such as algol 68 [VANW75], Pascal [WIRT71] and Ada [ICHB83]. It should be noted that this type of binding is conventionally made up of two parts, static scoping and static type checking. A variation is to allow static scoping of names but to check the types of objects dynamically. This method of binding is used in the language SASL[TURN79] and provides a measure of polymorphism.

In dynamic binding the association between name and object is made during the dynamic evaluation of the program. For example

```
let  a := 3
let  c := proc( )
          begin
                . . .
                a := -2
          end
c( )                               ! a contains 3 before and -2 after
begin
      let  a := -19
      . . .
      c( )                         ! a contains -19 before and -2 after
end
```

Dynamic binding is made up of dynamic scoping with either static or dynamic type checking. For static type checking with dynamic scoping all objects with the same name in the program must have the same type. It is much more common to find dynamic scoping linked with dynamic type checking as first provided by LISP [MCAR62]. For the rest of this paper, unless explicitly stated we will use the term static binding to mean static scoping with static type checking and dynamic binding to mean dynamic scoping with dynamic type checking.

The advantage of statically bound programs is that some errors may be detected early in the software life cycle. This is so important to some language designers such as those of ML [MILN84], Argus [LHJSW83] and Pebble [BL84] that they insist that all binding be static. Where it is possible to perform static checking it is also possible to make the programs more efficient since checks can be factored out.

In large scale systems the issue of binding is of major importance and it is in such systems that static binding has hidden costs. Any change to program or data form requires a recompilation to re-establish the bindings. For large systems involving software evolution or the distribution of software this recompilation may be prohibitive in cost.

It is interesting to note that to overcome these costs and to accommodate dynamic evolution of programs and data, most programming languages allow file names to be bound dynamically and read statements to involve a dynamic type check. This is true even in, so called, strongly typed languages such as Pascal. Indeed it is not well recognised that these languages allow static binding within the program and resort to the operating system, file system or database management system to provide dynamic binding. The two binding mechanisms in these languages almost never obey the Principle of Correspondence completely.

In the context of persistent systems there is a requirement for both types of bindings. A static binding would interpret the above file names in the compilation environment, thus freezing the program and data form. The second type of binding would interpret the file names in the context of the run time environment. This is dynamic binding.

We assert that both methods of binding are necessary for large scale system construction and evolution. We recognise that it is advantageous to statically bind wherever possible with dynamic binding available where necessary. We would expect small objects to be statically bound to form larger objects, with interfaces between those objects being dynamically bound. We coin the term flexible incremental binding (FIB) to describe this mixture of bindings, which we expect to obey the Principle of Correspondence.

3. A MECHANISM FOR CONTROLLED DYNAMIC BINDING

In Section 2 the case for some dynamic binding (and consequently dynamic type checking) was made. In this section we introduce name spaces, an environment mechanism that permits the following:

(i). the storage of bindings in a name space;

(ii). the dynamic use of names from a name space;

(iii). the static use of names from a name space;

(iv). the evolution of the names available in a name space;

(v). safe exchange of arbitrary data between parts of the system - especially between the permanent store and programs.

Remember that a name may be statically or dynamically bound, and the system may determine the type and object associated with the name. However, we wish to compile most code in contexts where there is enough information to statically bind and statically type check. Even when dynamic binding or checking the strictness of type checking will not be compromised.

3.1 Abstract name spaces

A name space expression is written

> **ns** identifier-list **from**
>   sequence
> **end from**

In such an expression, declarations for each of the identifiers in the identifier-list will appear in the sequence after **from**, giving each identifier a type and an associated object. For example

```
let  an1 = proc( x : int ; y : string -> ns )
           ns a,b,c from
               let xx = x * x                ! let is an initialising
               let yy = y ++ y               ! declaration in Napier,
               let a = 3                     ! with type inferred from
               let b:= xx - 2 * x + 5        ! the expression
               rec c = proc(i:int->string)   ! recursive declaration
                   if i<= 0 then yy
                   else c( i - 1 ) ++ c( i - 1 )
           end   from
```

All instances of the name space yielded by this procedure start off by containing three triples as shown in the following table

identifier	associated object's type	associated object's value
a	int	3
b	var int	a location initialised to x^2-2x+5
c	proc (int -> string)	a procedure to yield 2^n doublings of the given string

Locations a and c are non updatable, ie constant. We show later the similarity between these name spaces and abstract data types, records and structure classes.

3.2 Instances of name spaces

To form an instance of a name space we provide parameters to the procedure.

For example:

```
let n1 = an1( 3,"to" )
```

This would yield a name space which presents the environment

```
{ a : int                    : 3,
  b : var int                : var containing 8,
  c : proc( int -> string )  : a proc which will give 2n doublings
                               of "to" }
```

8

3.3 Presenting environments

When we write

```
let sequence1 in sequence2
```

in languages such as ML, *sequence2* is compiled in the context of the environment yielded by *sequence1*.

When we write

```
proc( parameters )
      statement
```

the 'statement' here is compiled in the context of an environment specified by 'parameters' but whose values are provided by the 'actual' parameters at the time of the call.

A name space provides an environment in a similar way, and the values provided are those left in the name space when it was created, or when it was last used. There are two constructs, one in which the name space instance used is statically determined and the other where it is dynamically determined.

3.4 Nomenclature

The nomenclature used is analogous to that for procedure parameters. We consider name spaces to be an abstraction of store of arbitrary permanence. The name space information at the time a code sequence is compiled constitutes a formal store and that used at the time it is executed is the actual store. The relationship between the formal and actual store differs in the above two constructs.

3.5 Static invocation of name spaces

A name space may be used to collect values, for example, left by previous runs of this and other programs. If it is known when a program is written which name space will be used, then an expression may be written which always yields the same result, and which can be evaluated both at compile time and run time, to identify that name space. The compiler can then be instructed to compile a sequence of code in the context of that name space's provided environment. For example

```
with n1 compile
     sequence
end compile
```

In the sequence, the identifiers a, b and c will be available with type and associated object as specified in 3.2 above. If, in the sequence, there was the statement

```
b := b + 1
```

then successive uses of this program would find new values in b, though b is still associated with the same object.

```
with expression compile statement
```

will be shown later to be a general form allowing the environments yielded by ADTs (which also provide records and structures) to be used. Indeed, the expression may also take the form of a void sequence containing declarations, in which case it is a denotation of a static environment in which to evaluate the statement or sequence after **compile**. In that case, and when a name space is statically bound, the statement after **compile** is always executed using the same actual store determined by the expression before **compile**.

With such static binding to name spaces we can model the stored workspace mechanism, called "persistence", in Poly [MATT85a].

3.6 Dynamic binding of name spaces

A statement of the form

```
using expression with signature compile
        sequence
end compile
```

indicates dynamic binding is required, then a check is made at compile time that the expression will yield a name space. The sequence is compiled with the environment of names and types specified by the signature. Thus the signature specifies a formal store, each time the statement is executed, the expression provides an actual store, possibly different on each occasion. It is therefore necessary to check that the store on this occasion holds data with the names and types specified by the signature. Thus the dynamic binding and checking is localised to the **using . . . compile** construct.

3.7 Data evolution

The raison d'être for name spaces is the need for a mechanism to accommodate and localise change.

A name space may be changed by the entry of new bindings and the withdrawal of old ones. For example

```
extend expression with identifier-list from
        sequence
end from
```

requires that expression yields a name space, and updates this name space **in situ**, adding the name: type: object triples identified and initialised by the code from **with** to **end from**. Similarly,

```
drop identifier-list from expression
```

removes all the definitions associated with the names given in the identifier-list from the name space yielded by the expression.

In order that a check which has been applied before a **compile** cannot become invalid during the sequence that follows, a lock is established on a name space for the duration of the sequence which prevents **drop** operations on that name space.

3.8 More about checking

When a name space is used statically it is necessary to check that it has properties required of it. To do this, each name space keeps a count of the **drop** operations applied to it. At the time of compilation of **with** expression **compile**, the **drop** count value is recorded in the code, at execution time, the current **drop** count is checked to ensure that no significant changes have take place. The compiler has to access the name spaces at compile time to obtain the names and types from it. The **drop** count check ensures the changes to the name space (its address is embedded in the compiled code) have not invalidated the original conditions.

Both dynamic and static binding constructs therefore only check on the names they use, consequently both mechanisms permit a name space to be extended without impact on existing programs.

In the dynamic case, the object bound to a name may change - either because of use of a different name space as the actual store or because of a **drop** followed by an **extend**. It is intended that the reuse of the same name with the same type is an indication that the same meaning and use is implied. Consequently the dynamic binding allows replacement of components, as does assignment to a variable object.

3.9 Assignment and equality

Assignment of name spaces is reference assignment and their equality is an identity test.

3.10 Illustrating the uses of name spaces

The intent of our languages is to accommodate more of the programmer's work within a single language. A few examples will illustrate this.

Two examples of name spaces which do not necessarily involve persistence are presented first. To call the compiler at run time in order to obtain executable results, type definitions etc. presents problems in typing the parameters of the procedure *compile*. With name spaces we can type it

```
proc( env : seq of ns ; source : sourcetype ; results : ns )
```

The sequence of name spaces identifies the context of the compilation, and the results are returned by extending the name space called *results*.

Since any programmer calling *compile* will have conventions on the names used and the types of objects generated, the results of the compilation can be extracted by a dynamic name space binding and subsequently used statically.

In code like the *compile* procedure above or in a general purpose index, data input, query or data output mechanism, the programmer will wish to report to the invoking code errors detected. We assume that these errors will be reported via an exception mechanism which conveys information via parameters. In the case of these general purpose applications the packing of the culprits into a name space allows culprits of any type to be conveyed back to the invoker.

In languages currently in widespread use, dynamic association between data and program is achieved by using names (usually as character strings) to access data in a file catalogue (or directory) maintained by the operating system. A name space may serve in this role, allowing any data to be named externally to the program, but now requiring that it is named and typed consistently.

As a name space is itself an object in the language, a name in a name space may be bound to a name space. Hence hierachical naming structures similar to directory hierarchies are possible.

Program libraries are obtained by putting procedures or abstract data types in name spaces. The programmer can choose whether to statically bind to a routine in the library and thus find that certain library maintenance operations result in the program needing recompilation, or to bind dynamically so that the program will operate as long as the library provides a routine of the expected name and type. The type checking, binding and loading for all data objects, which have been described and prototyped[AM84], replace such things as: type checking linkers, consolidators, APSE databases and associated utilities. Separate compilation is achieved by running programs which leave program parts in such a program library.

Many operating systems identify a home directory for each user, where a user is identified by a string. This mechanism can be obtained by building a relation (or similar associative structure) which given a string returns a name space. Of course, a name space itself would be an adequate associative data structure if we use names to identify users.

Permanent storage is usually provided by the filing system of an operating system. Programs use this storage to convey results to later programs. Even applicative languages, and those that save workspaces, perform updates on such a store. By distinguishing one or more name spaces, say PermanentStore etc, which for a new system starts empty, and arranging to preserve all data reachable from them, as we have explored in PS-algol [PS85], then permanent storage is provided and the interprogram communication previously provided by the file system may now be coded within the language. Furthermore different users may adopt different naming schemes and organisations of permanent store.

A common requirement is to name a complex data structure, so that an instance of it can be identified for an activity with a given program. For example, a CAD system may be used for many projects. The data associated with each project may be a name space, and all of these may be named in PermanentStore. The current programs constituting the CAD system may be in another name space, say 'CAD', also made accessible by being bound in PermanentStore. The user may apply any program (procedure) from CAD to any of the project databases. The vendor of the CAD system may provide replacement versions from time-to-time, which will be installed, replacing the previous value of 'CAD'. They can be written so as to bind to the existing projects, but so as to extend them to contain new types of data. Bindings via name spaces of customised programs the package user has written will still run. Thus revised software or other data may be delivered to a customer without revealing the program source or proprietary data structures, and this can be used with existing data and software, which may represent a substantial investment by the customer. Such dynamic reassembly is essential with large commercial systems, and is not provided by systems dependent only on static binding, when they claim to provide persistence.

3.11 A comparison with other systems

As has already been remarked, the static binding of name spaces, achieves the same effect as the statically bound work spaces, such as those in Poly. Pascal/R [SCHM77] has a permanent type constructor **database** which provides an environment which is a binding of names to relational values. Only one is permitted per program, and its type is defined as for all types in Pascal, so that the program is compiled in the statically determined formal store. At execution time the actual store is determined by a parameter to the program, and is, presumably, dynamically checked to be of a compatible type. Transient instances of **database** cannot occur, it cannot be used in other structures and its contents are limited in type.

In Galileo [AOO82] **use** statements add bindings to a standard environment which is permanent. **enter** statements arrange for subsequent environments to be compiled in the identified environment. This is equivalent to writing an

```
extend PermanentStore with name from declaration
```

and writing

```
with PermanentStore compile statement
```

respectively. Galileo does not appear to accommodate any form of dynamic binding, but as environments may be held in structures, and may hold data of any type, it achieves the same power of structured naming as the static name space mechanism we describe does.

In Amber [CARD85], some of the effects of dynamic binding are achieved using inheritance. As the dynamic binding of name spaces only checks the names in the signature the mechanism proposed offers similar inheritance. **extend**ing a name space then becomes equivalent to specialisation.

In Taxis [MW80], all structures are permanent and one statically bound implied name space seems to be the underlying model.

In Adaplex [SFL83], each database is restricted to hold only entities and is statically bound to the programs that use it. It is always persistent. With a type declaration corresponding to each entity type, a name space corresponding to each database, and using only static binding of name spaces, the persistence of Adaplex could be modelled by the name spaces proposed here.

Pebble is also intended to facilitate the assembly of large systems, but is based on the assumption that dynamic binding is not required. The triples stored in a Pebble environment seem to be the same (except for details) as those in our name spaces. However, we have proposed additional operations and an additional form of binding, to permit system evolution in large systems.

3.12 Outstanding problems

With file directories it is possible to write general utilities which scan a directory, often interactively, revealing the directory's contents and possibly allowing user action - e.g. delete - on each file. Some equivalent operations on name spaces may be useful, however, it is difficult to identify a primitive in terms of which they may be written. For example, how does one type the control variable of an iteration over a name space and how does one represent and manipulate names? This problem is not peculiar to name spaces, it also arises in scanning other structures that yield an environment when writing general utilities for browsing, diagnosis, statistics collection and generalised data input and output. Current practice is to step outside the type system or to depend on calling the compiler. Our desire to support more of the total programming activity within the strictly typed language leaves us confronting this problem.

4. TYPES AND DATA STRUCTURES

The data types in Napier are defined by rules of the following form:

a. The primitive data types are integer, real, boolean, picture, pixel, and string;

b. image is the type of an image made up of pixels arranged as a rectangular matrix;

c. For any data type T, *T is the data type of a vector with elements of type T;

d. For any data types $T_1,...T_n$ and T, **proc**$(T_1,...T_n->T)$ is the data type of a procedure taking parameters of type T_1 to T_n and producing a result of type T. The type of a similar resultless procedure is **proc**$(T_1,...T_n)$;

e. For any data types $t_1,...t_n,...t_m$ and identifiers $id_{n+1},...id_m$, $id(t_1,...t_n)\{id_{n+1}:t_{n+1},...id_m:t_m\}$ is the type of a user defined type with parameters $t_1,...t_n$ and signature $\{id_{n+1}:t_{n+1},...id_m:t_m\}$ and type name id;

f. For any data types $t_1,...t_n$, $t_1|t_2|...|t_n$ is the type of a union of the types $t_1,...t_n$;

g. **ns** is the type of a name space;

h. For any data types $t_1,...t_n,...t_m$ and identifiers $id_1,...id_n,...id_m$ then **rel**$(id_1:t_1,...id_n:t_n->id_{n+1}:t_{n+1},...id_m:t_m)$ is the type of a relation with index $(id_1:t_1,...id_n:t_n)$ and result $(id_{n+1}:t_{n+1},...id_m:t_m)$.

The universe of discourse is defined by the closure of rules (a) and (b) under the recursive application of (c), (d), (e), (f), (g) and (h).

A type in Napier is an abstraction over a declaration just as a function is an abstraction over an expression and a procedure an abstraction over a statement. The main difference between these abstractions in a language with higher order functions is in the use of names. A type definition introduces names into the environment which may then be used to access values of the type. The type name itself may also be used in subsequent declarations. This type mechanism allows concepts such as records, structures and abstract data types to be captured in one. For example, a Pascal record may be modelled by

```
let complex = type( rpart, ipart : real ) is rpart, ipart

let add = proc( a,b : complex -> complex )
            complex( a_rpart + b_rpart, a_ipart + b_ipart )

let a = complex( 3.1,2.1 ; let b = complex( 3.1,-4.1 )

let c = add( a,b )
```

In the above, the type declaration introduces the name *complex* which can be used as a type identifier and as the type constructor. Objects of the type may be created by passing parameters to the constructor. For example

```
complex( 3.0,-42.1 )
```

creates an object of the type *complex* defined above. A value of the type can be obtained using the underbar notation. Thus

```
a_rpart
```

yields the *rpart* value of the complex object *a*. Notice that the representation of the type is not hidden in this example.

Each type has a signature that is used in implementing both polymorphism and name spaces. The above type has the signature {*rpart*: : **real**,*ipart* : **real**}. The type is parameterised in the same manner as procedures.

The mechanism may also be used to implement abstract data types. For example, to hide the representation of complex we could write

```
let compTYPE = type() is add,constr from
    let complex = type( rpart,ipart : real ) is rpart,ipart
    let add = proc( a,b : complex -> complex )
            complex( a_rpart + b_rpart,a_ipart + b_ipart )
    let constr = proc( a,b : real -> complex )
            complex( a,b )
    end    from

let mycomplex = compTYPE()
let add = mycomplex_add ; let create = mycomplex_constr
let a = create( 3.0,2.0 ) ; let b = create( 3.1,-4.2 )
let c = add( a,b )
```

Using the normal algol scope rules, the definition and therefore the selectors on *complex* are hidden to the outside world. Only the primitive functions *add* and *constr* are exported.

4.1 Type matching

One major problem that arises in any type system is the meaning of the equality of types. This problem is made more difficult when the language allows types to be stored and reused as is the case in Napier. As an example of these difficulties, we may write

```
let newtype = if . . . then one type definition
            else another type definition
```

That is, the *if* clause may return a type, as may a procedure or a type itself and we cannot determine which one at compile time. Russell and Poly have different solutions to this problem and Napier proposes a third. A type is a data object in Napier which may therefore be named and stored like all other data objects. A type has a parameter list which is used to create objects of the type and a signature which specifies the set of values for the type. Two expressions which yield types are only type compatible if the types have the same parameter list in one to one correspondence and the same signature. Thus the above *if* expression is only valid if the two type expressions are compatible.

Type names may be declared and used as a shorthand in parameter lists and signatures. This is the most convenient way of using types but raises other problems in type checking in languages with higher order procedures. We would like objects with the same type name to have the same type. However consider the following program

```
let  c = begin
                let complex = type( b : Int ) is b
                let knit = proc( a : complex )
                        . . .
                knit
        end
let ccomplex = type( rpart, ipart : real ) is rpart, ipart
let  a = complex( 1.5,-2.3 )
c( a )          ! this is an illegal call of c
```

Clearly the call of *c* is illegal since it is really the procedure *knit* which refers to a different *complex* . To overcome this problem, two objects only have the same type if the types have the same name, the parameter lists are in one to one correspondence and they have the same signature. This allows types to have different internal properties but still be regarded as the same. The significance of this is clear to anyone who has tried to evolve persistent data using both old and new versions simultaneously.

In Napier, there is a union of all types called **type**. This is used when storing types in a name space or passing them as a parameter. All types match the type **type**. However as we will see the type may be restricted by a signature to allow a more specific partial match for a particular type.

5. DISCRIMINATED UNIONS

To define recursive data structures we use a discriminated union. For example to build a list of integers we may use

```
rec intlist = union type() |
              type( hd : int ; tl : intlist ) is hd,tl
```

The list may be made up of pairs or an empty value. The parameter lists of the constituents of the union must be unique in order to distinguish them on initialisation. Alternatively they may be named. To construct and print a list we could use

```
let nil = intlist()

let printlist = proc( head : intlist )
                while head ~= nil do
                begin
                        write head_hd
                        head := head_tl
                end

let head := intlist( 1,nil ) ; head := intlist( 2,head )
printlist( head )
```

It is interesting to note that projection from a union always carries a run time check. In the above case we must always check that the selector is appropriate to the object. The check can be factored by a project operation. For example

```
rec scalar    = type( name : string ) is name &
    cons      = union  type() |
                type( hd : comptype ; tl : cons ) is hd,tl
    PROC      = type( args : cons ; t : comptype ) is args,result &
    comptype = union scalar | cons | PROC

let nil = cons()

let eq = proc( a,b : comptype -> bool )
         project
         a onto scalar and b onto scalar : a_name = b_name
         a onto PROC and b onto PROC : eq( a_args,b_args ) and
                                       eq( a_result,b_result )

         a onto cons and b onto cons : ( a = nil and b = nil ) or
                                       a ~= nil and b ~= nil and
                                       eq( a_hd,b_hd ) and
                                       eq( a_tl,b_tl ) )
         default                     : false
         end  project
```

In the above *scalar* and *PROC* case the type checks can be factored out at compile time. However the other projection only projects onto another union and a further projection is required to factor out all of the checks. The patterns of ML could be used here.

6. POLYMORPHISM

The method of defining polymorphic functions in Napier is taken directly from Russell [DD79] and Poly. It is parametric dependent types. For example, a polymorphic sort routine could be defined by

```
let sort = proc[ t : type() with {less :proc(t,t ->bool)} ](A :*t)
           ! Your favourite sort routine using less
```

The first parameter *t* is a type which, since it is enclosed in square brackets may be implied from the other parameters at the time of call. The type *t* must have at least a function *less* defined in its signature which takes two objects of type *t* and returns a boolean. That is, we only require a partial match between signatures of the actual and formal type parameters. The second parameter *A* is a vector of objects of type *t* . To call the procedure we might use

```
sort( int with '<' as less, a vector of integers )
```

There is a little renaming here to allow the integer infix operation '<' to be used as *less* inside the procedure. Of course, we can partially apply the procedure to freeze the type. Thus

```
let int.sort = sort( int with '<' as less )
```

defines a function *int.sort* that may be applied to vectors of integers.

A more exciting prospect is the possibility of types themselves being parameterised by polymorphic values. For example, all pairs of values of the same type may by defined by

```
let  all.pairs = type[ t :type() ]( first,second :t ) is first,second
```

Then all integer pairs could be defined by

```
    let  int.pairs = all.pairs( int )
```

All of the lists of elements of the same type can be defined by

```
    let  list = union[ t :type() ]type() |
               type( hd : t ; tl : list( t ) ) is hd,tl
```

Thus

```
    let  int.list = list( int )
```

defines the type of a list of integers. We can also define lists of mixed elements by not parameterising the second element of the list. For example

```
    let  list = union[ t : type() ] type() |
               type( hd : t ; tl : list( t ) ) is hd,tl
    let  nil = list( int )()
    let  intelement = list( int,3,nil )
    let  mixed.list = list( string,"ronald",intelement )
```

7. A DATA TYPE COMPLETE FORM OF RELATIONS

Relational constructs in a language are well known [SCHM78] [WSKRV81] [ROWE80] [SHOP79] and other persistent languages accommodate classes of entities [AOO82] [NAUR63] [SFL83]. The relational languages, however, typically restrict the types over which the relation may be formed, presumably to reduce implementation problems or to adhere to first normal form (1NF) [CODD70]. The notion of an atomic value, on which 1NF is based, is however relative; most such languages allow character strings as a column type, and also have a substring operation, for example. Consistent with our adherence with the principle of data type completeness, we allow a column to take any value permitted in the language.

7.1 Relation types

We intend the programmer to visualise the relation as a table, as a sparse array or as an extentionally defined partial function. For this reason, the type declaration is similar to that of a procedure i.e.

```
    rel( parameters  -> parameters )
```

The names and types in the first parameter list identify the columns that form the key of the relation. The names and types of the second parameter list identify the columns dependent on the key. Some examples of relation type declarations follow:

```
let rel1 = rel( i : int -> s : string )

let rel2 = rel( ai : *int -> x : *pixel ;
                y : proc( int -> rel2 ) )
```

A denotation for relational constants is achieved by using the type name, or type expression as a relational constructor. For example

```
let r1 = rel1{
              ! i    ->    s
              0      ->    "zero"
              1      ->    "one"
              2      ->    "two" }
```

7.2 Applicative relation operations

The usual relational algebra operations join, project, select, union, intersection and difference are defined over objects of this type. These are applicative, so they generate new relational instances, whose type can be deduced from the operands and operator.

7.3 Transfer operations

A difficulty facing the language designer for a language with bulk data structures is the design of an interface mechanism. We use indexing, an iterator and a de-setting operator.

Indexing is written exactly like function application or array subscripting. For example

```
r1( 2 )
```

would yield the 'result' part of the tuple with index value 2 in relation r1. In general a sequence of expressions provides values for an equality match on the index columns (those appearing before the arrow). If there is no match, an exception is raised. If there is a match then the result is an actual store, selected by this index, with the environment corresponding to the signature appearing after the arrow in the relation type definition defining the formal store for compilation. Thus it may be used exactly like an ADT, both in the LHS and RHS contexts. As the index environment does not appear in this store, the semantic and engineering problems of update to an index are avoided.

The iteration is written

```
for each x in y do statement
```

This is a general form applicable to arrays and relations. The variable x takes all the values in turn, in some arbitrary but repeatable order, corresponding to the components of y . In the case of a relation it takes the value of an ADT with environment formed by the concatenation of the index and result signatures. The statement is compiled in the context of a formal store defined by this environment, with the index fields protected as constant, to prevent update to the index. The *statement* is executed once with each actual store corresponding to a tuple in the relation.

De-setting is written

```
the expression
```

The expression has to have a relational type, and typically it is a selection expression e.g.

```
the  rl where s = "one"
```

If the value of the expression is not a singleton set then an exception is raised. If it is, then an actual store corresponding to the tuple, with index fields protected against update, is the result of the whole construct. It may be used in a LHS or RHS context.

The transfer interface in all three cases is based on a statically typed, but dynamically selected and bound store substitution. Thus it provides for individual access and update operations on the bulk data.

7.4 Bulk update operations

In situ update of bulk data is deemed necessary for semantic (data sharing) and engineering reasons. In situ update via the transfer operations above already exists. Expressive power and engineering advantage can be gained by providing bulk update statements. These are

```
update x adding y

update x removing y

update x substituting y
```

The effect of the first is to make x now have the union of the tuples in x and y. The effect of the second is to remove all tuples from x that have matching tuples in y. The third substitutes for columns in the result of x new values from y for the types whose indexes match in x and y.

7.5 Avoiding a semantic anomaly

There is a well known update anomaly, to do with updates altering selection criteria, typified by discussion about updates such as

"fire all managers who earn less than 1.5 times
the average salary of their immediate subordinates"

We take the view that these semantic ambiguities are best removed by making the code equivalent to identifying all the updates with the data unchanged, and then performing the updates. It is supported in the language by a dynamic lock on relations, so that, while they are the subject of a selection or iteration scan, no in situ updates may be performed.

8. CONCLUSION

Name spaces have been proposed as a mechanism to formalise the association between stores and program. They are motivated by our observations on longer term storage. Static binding mechanisms are supported, which we see as sufficient for small systems and short term data. A dynamic mechanism for binding actual stores to formal stores is proposed to accommodate the changes inevitable in longer term systems which cannot be economically dealt with by total rebuilding in large systems.

The type relation has been presented as a bulk data representation, with the usual restrictions relaxed in favour of data type completeness. The abstract data type, union and parametric polymorphic types are presented to allow precise data description. The combination of these with the type and binding rules for name spaces is shown to be simple and consistent.

The model proposed enables the whole program to be statically bound and typed except for well identified points which may be introduced by the programmer if the flexibility is required. Even when the dynamic construct is used the type checking, though delayed, is no less strict.

REFERENCES

[AM84] [AB85] [ABCCM81] [ABCCM83] [AOO82] [BL84] [CARD85] [DD79] [LAND66] [LHJSW83] [MATT85a] [MCAR62] [MILN84] [MW80] [NAUR63] [PS85] [ROWE80] [SCHM77] [SCHM78] [SFL83] [SHOP79] [TENN77] [TURN79] [VANW75] [WIRT71] [WSKRV81]

Chapter 2

Modules and Persistence in Standard ML

Robert Harper
University of Edinburgh

1. INTRODUCTION

Standard ML [MILN85] is an interactive, statically-typed programming language that provides support for higher-order functions, user-defined abstract types, and type-safe exceptions. The ML modules facility [MACQ85] provides the means for organizing ML programs into units, called *structures*.. Each structure has a well-defined interface, called its *signature*, that plays a role similar to that of types in the core language. Viewed statically, a program is a hierarchical arrangement of interdependent structures. Support for the dynamics of program construction is provided primarily by *functors*, functions from structures to structures, that provide the means of glueing structures together to form coherent units. By "coherent" we mean that the individual structures comprising the program are combined so as to share information in the intended way.

A distinctive characteristic of the ML modules system is the way in which this sharing between structures is managed. We use the word "sharing" to refer to any of several ways that two structures can come to depend on one another. There are, in essence, two ways in which dependency can arise. If a structure defines a type that is used by another structure in an essential way (say, by defining a function over that type), then the second structure can only be used in an environment in which the first is present as well. A similar form of dependency arises when one structure allocates a variable or defines an exception that is used by a function in another structure. In this case the two usually must be treated as a unit, for if the first structure is recompiled, it generates a *new* location on the heap (or new exception) that cannot be accessed by the second. MacQueen's central observation is that both forms of dependency can be reduced to managing the sharing of common substructures. In particular, functors which are used to build up hierarchies of structures may require, as part of their parameter specification, that two structures be built from the same structure, thereby ensuring that the combination is properly coherent.

These mechanisms appear to form an adequate basis for addressing the theoretical problems that arise in a modular programming environment. In this paper we address some of the more pragmatic issues involved, such as separate compilation, preservation of session context, and program libraries. These issues are traditionally taken to lie outside of the boundaries of the programming language itself, but this seems to be inappropriate for an interactive langauge like ML. Furthermore, recent work [ACC81] [AM85b] [AGO85] has shown that considerable benefits accrue from integrating these "environmental" issues into the language proper. The key idea that we shall adapt from this work is the idea of a *persistent heap*, whereby the program heap is considered to be a permanent, rather than transient, repository of objects used by a program. Objects are accessed by a generalized address, called a *persistent identifier*, or *PID*.

Facilities for separate compilation and so forth can be naturally viewed as special cases of persistence. Our claim is that although a persistent heap is adequate for our purposes, it does not appear to be necessary. In particular, we show that the sorts of facilities that are most commonly used in a program development environment can be implemented by techniques that, while not as general as a full persistent heap, are considerably simpler. At the same time, however, we demonstrate that viewing these problems from the point of view of persistence is both natural and fruitful.

2. THE ML MODULE FACILITY

We begin with an overview of the ML module facility [MACQ85]. Readers familiar with [MACQ85] or [HARP86] may safely skip to Section 3.

A structure is an encapsulated environment. The basic form of structure expression consists of a declaration between the brackets struct and end. The declaration may define new types, exceptions, variables, or structures. A structure declared within another structure is called a *substructure* of that structure. Substructures are used to express hierarchical dependency relationships among structures, and are an essential aspect of ML's approach to sharing. The declaration embodying a structure may contain non-local references only to previously-defined structures. This ensures that a program consists *only* of a collection of structures: no stray top-level declarations that are not packaged into a structure are allowed to participate in the construction of a program unit.

Here are some examples of structure declarations:

```
structure R =
  struct
    datatype t = ...
    val f : t -> t = ...
  end

structure S =
  struct
    structure R = R
    datatype 'a tree = leaf of 'a | node of 'a tree * 'a tree
    exception nonexistent : R.t
    fun search(t:'a tree,x:'a) = ...
  end
```

The keyword structure introduces structure bindings, much as val introduces a value binding. The structure bound to R defines a new type t and a function f on that type. The structure bound to S imports R as a substructure, and defines a new type 'a tree, an exception nonexistent, and a function search. Notice that the type t defined in the substructure R of S is accessed by a *qualified name*, consisting of a structure name and a component name, separated by a dot.

Hierarchical dependency of one structure on another is expressed by the use of substructures. By binding the structure R in as a substructure, the structure S makes its dependence on R explicit, and furthermore makes S self-contained. Were R not incorporated as a substructure of S, the dependency of S on R would be implicit, reflected only in the fact that the type of the exception nonexistent refers to R, and not by the existence of any explicit reference to R within S. Such a situation is undesirable because S can only be meaningfully used in a context in which R is available as well, and if one does not tie the two together in some way, then an essential coherence in the structure of the program is lost.

Structure sharing occurs when two structures incorporate the same structure as a substructure. For example, if some structure Q were to bind in R as a substructure, then S and Q would be related by virtue of the fact that they both depend explicitly on the structure R. It is crucial that both S and Q have the *same* instance of R as substructure, for otherwise coherence is violated. The reason, in this case, is that datatype declarations in ML are *generative* in the sense that each elaboration of a datatype declaration defines a new type. Therefore if S and Q do not refer to the same instance of R, their components will not have compatible types. A similar problem would arise if R declared an exception that is handled or raised by functions in S and Q.

It may strike the reader as peculiar that we are emphasizing the need to maintain version control, especially in the setting that we have developed so far. At present we are describing a static configuration of structures, but later on we will turn to the problem of generating and updating these configurations. From that point of view, preservation of sharing is a crucial problem.

One of MacQueen's central observations is that the problem of dependency between structures can be reduced to preserving appropriate sharing relationships between structures and types. In this paper we will restrict our attention to structure sharing. There is no loss of generality in this assumption (a type can be taken to be a structure consisting of only the type declaration), but in practice this is inconvenient, and the full modules system treats type sharing separately from structure sharing. The reader is referred to [CM85] for a discussion of type sharing in a persistent environment.

The key to managing structure sharing is to have a suitable notion of equality between structures that is sufficiently fine-grained as to capture the informal idea of "versions" or "instances" of a structure during program development. Each structure expression, when elaborated by the compiler, generates a new structure that is distinct from all other structures previously declared. This generalizes the generativity of data types mentioned above, and also captures the generative character of allocations of references to the heap. Structure equality is then defined to be in terms of a unique identifier (address or timestamp) associated with the structure. This identifier is used as the basis for structure equality tests in functors.

Signatures describe the interface of a structure, describing the names and roles of the identifiers declared within a structure. This information consists of the names of the type constructors defined in the structure, the types of the variables and exceptions, and the signatures of the substructures. Signatures are expressed by writing a *specification* between the brackets sig and end. Signature expressions are subject to an even more restrictive closure rule than are structures: they may contain no references to any external identifiers other than the pervasive primitives. They may, of course, be bound to identifiers using a signature binding, as follows:

```
signature SIGR =
  sig
    type t
    val f : t -> t
  end

signature SIGS =
  sig
    structure R : SIGR
    type 'a tree
    val search : 'a tree * 'a -> bool
  end
```

The utility of signatures lies in the relation called *signature matching* between structures and signatures. Roughly speaking, if a structure satisfies the specification given in a signature, then that structure is said to *match* that signature. The exact definition of signature matching is somewhat complex, but the idea is quite intuitive. For example, the structure S defined above matches the signature SIGS because

1) The structure S.R matches SIGR;
2) The type S.tree is a one-argument type constructor;
3) The variable S.search has the type given in SIGS.

This rough idea of the exact definition of signature matching suffices for our purposes. The interested reader is referred to [MACQ85] [HARP86] for a more precise account.

A signature SIG is said to be *more general than* another SIG' if every structure that matches SIG' also matches SIG. For example, the signature SIGS' obtained by deleting the specification for search from SIGS is more general than SIGS. This is derived from the fact that signature matching is liberal in the sense that the structure may have "extra" components that are not specified in the signature.

A signature may be (optionally) attached to the structure identifier in a structure declaration, indicating that the structure on the right of the binding must match the given signature. For instance, we may qualify the declaration of structure S above with the signature SIGS as follows:

```
structure S : SIGS = ... as above ...
```

Since the structure bound to S matches the signature SIGS, this is an acceptable declaration.

A program consists of a set of interrelated structures, encapsulated into a single root structure that binds the collection together into a unit. Each structure presents a well-defined interface to the others, and sharing is controlled by substructure bindings. Viewed statically, this appears to capture the idea of modular programming. But from the point of view of the dynamics of program development, we are lacking sufficient tools to create and modify such an arrangement of structures. For example, if we wish to modify the code of a function in the structure R above, then we need to recompile it and rebuild structures S and Q, using the same instance of R in both cases. It is impractical to insist that we simply recompile the entire program, thereby recreating the static configuration outlined above. What is necessary, of course, is some means of *relinking* S and Q with the new copy of R. This facility is provided by functors.

A functor is a function that, given one or more structures as arguments, yields a structure as result. The main use of a functor is to build a composite structure from argument structures, as we have indicated. Consider the structures R and S defined above. In order for S to be able to sustain a change to R, it must be possible to rebuild S with a new copy of R replacing the old copy. To do this, one defines a functor FS that, given a structure with signature SIGR, yields a structure with signature SIGS. Structure S is built by applying FS to R, yielding a new version of S with the new R incorporated as substructure. Similarly, the structure R itself is generated by a functor, but since R lies at the bottom of the dependency hierarchy, its functor has no parameters.[1]

```
functor FR() : SIGR =
  struct
    datatype t = ...
    val f : t -> t = ...
  end

structure R : SIGR = FR()

functor FS( R : SIGR ) : SIGS =
  struct
    structure R : SIGR = R
    datatype 'a tree = ...
    exception nonexistent : R.t
    val search(t:'a tree,x:'a) = ...
  end

structure S : SIGS = FS( R )
```

The dependence of S on R is flexible because S is built by applying FS to R. Since FS uses R abstractly (knowing only its signature), any instance of R can be plugged in to S by applying FS to that instance.

Now a functor may contain a sharing specification to ensure that two structures are built from a common substructure. For example, consider the structures S and Q discussed above, each with an instance of R in common. Then a functor to build another structure, say W, from S and Q will usually wish to require that it be given *compatible versions* of S and Q, for otherwise the combination will not be coherent. This is achieved by the sharing specification, as follows:

```
functor F( S: SIGS, Q:SIGQ sharing S.R = Q.R ): SIGW = ...
```

An application of this functor to two structures is acceptable only if the substructures R of each argument are equal in the sense of having the same unique identifier or address.

We shall limit our attention in this paper to "pure" functors, those without non-local references to the environment. The main significance of this restriction is that functors need not be represented as closures, and certain problems related to the signature of an impure functor are avoided. Another consequence of this restriction is that the structure resulting from a functor application can depend only on the parameter structures, and not on any other structures in the environment.

---

1 The point of defining a nullary functor for R is mainly that in general one may want to obtain multiple instances of R without recompiling the body.

3. PERSISTENCE

In this section we discuss the problems of separate compilation, session preservation, and program libraries in the context of the ML modules facility. The essential observation is that each of these problems can be addressed from the point of view of persistent data. We will focus our attention on structures and functors as the units of persistence, as these are the fundamental units of program in ML. A more fine-grained notion of persistence would clearly be useful for some applications, but these issues lie beyond the scope of this paper. First, we discuss a general notion of persistence, and discuss its interaction with sharing and its suitability for support for program development. Then we isolate several special cases of the general notion, each motivated by a particular, limited application, but admitting a much simpler implementation than the most general form.

The most general notion of persistence, which we shall call *object persistence*, consists of viewing all objects as existing in persistent storage, with ephemeral storage serving only as a cache for quick access. Each object is identified by a *persistent identifier*, or *PID*, which is the address of that object in persistent storage. In effect, the program heap resides in persistent storage, with local caching in ephemeral storage for the sake of efficiency. The heap is garbage collected as usual, so that only accessible objects are preserved. In order to ensure that all accesses to persistent data are type safe, each object must have its type associated with it, and some sort of run-time typechecking is essential. When types are limited to pervasive primitives like int and bool, this is a well-understood notion. The situation becomes more interesting when user-defined abstract types are introduced; see [CM85] for a treatment of this problem.

It has proved fruitful to view a persistent heap as a generalized form of database, and therefore it is normal for a program to access several persistent heaps. This entails that persistent identifiers must be relativized to to a particular heap, and also that primitives be provided for accessing heaps. A program connects to an address space using a primitive called connect. Access to address spaces is mediated by a *handle* [CM85], typically a file name, that designates the address space to the run-time support system. Thus connect(H) returns some form of descriptor required for accessing the address space identified by the handle H. It is assumed that all changes to the persistent store are automatically preserved, and that any support for simultaneous access to an address space is provided by the underlying support primitives.

In the context of ML structures and functors are persistent objects, and therefore they are assigned PID's at creation time. The PID is used to mediate all access to the structure or functor, and it is also used to define equality between structures: two structures are equal if and only if they have the same PID. This corresponds to the fact that structure expressions are generative, in that each elaboration of a structure expression yields a unique structure. Functors are simililarly assigned PID's when they are created, but since there is no equality test for functors in ML, there is no need to be especially aware of this fact.

Notice that since we required that dependencies between structures be represented by a substructure hierarchy, it is always the case, even in a persistent environment, that related structures cannot be separated from one another. For example, since the structure S defined in Section 2 incorporates R as a substructure, the PID of R is bound to the local identifier R of S, thereby ensuring that the appropriate version of R is properly associated with S in the persistent heap.

It has been observed [CM85] that it is necessary in a persistent environment to associate the type of a structure with the object in persistent storage in order to ensure type safety. In the context of ML, this means that signatures must be stored with each persistent structure, and a pair of signatures (one for the parameter and one for the result) and the sharing specification must be stored with each functor.[2]

Object persistence is the most general form that we could implement, and therefore is adequate as the basis for supporting the needs of a program development environment mentioned above. Session preservation is completely automatic in such an environment since every heap is implicitly preserved and can be accessed with the connect primitive. Thus we can view a session with ML as being initiated by connecting to a persistent address space, doing some work, and then exiting. To return to that session at some later time, we simply reconnect to that address space, and resume where we left off.

Of course, if all we are interested in supporting is session peristence, then a much simpler implementation strategy is available: simply write a snapshot of the heap to persistent storage. A session is resumed by reloading the snapshot into a virgin ML system. This approach is rather crude and inefficient since we have no choice but to treat the entire heap as a monolithic object, without the possibility of treating any fragment of the heap separately, or being able to access it alongside other heaps.

One common application of session persistence is for a special case that we might call *program persistence*. The idea is to regard the significant result of a session to be a single structure consituting the root of some program hierarchy. For example, in the current implementation of ML, the ML compiler itself is merely a checkpointed heap that we reload in order to compile a program. The difficulty with this approach (aside from the aforementioned awkwardness of full checkpointing) is that it amalgamates the programmer's heap with the heap of the compiler itself. Thus if a user wishes to construct an ML program, the entire heap, including the compiler itself, must be written out as a unit.

Of course, if we were to use full object persistence, then there is a much better solution to the problem of program persistence. Since a program consists of a hierarchical arrangement of structures, each with a PID, then we can isolate a program as an independent address space by exploring the set of objects accessible from the root structure, and writing this out as a self-contained heap. Then any program may connect to this heap and use it freely, without having to replace the entire context as with simple-minded session persistence. Thus the problem of program persistence is solved by having persistent structures.

In fact, it is not necessary to have full object persistence in order to implement this strategy. All that is needed is the ability to collect together the accessible portion of the heap as a unit, and write this to persistent storage. Copying garbage collectors already provide this facility, since they work by consolidating the accessible portion of the heap in a contiguous region of virtual memory. If we define "accessible" to mean "accessible from a given structure", then we can use the collector to isolate that portion of the heap relevant to a given program, and write it to persistent storage. The main difficulty that arises is that one must be prepared to relocate this fragment of the heap when it is reloaded. But this is quite simple, using essentially the same technique in reverse. To reload a fragment of the heap, simply use the collector to shift the current heap in virtual memory so provide space for the saved heap, then read the saved heap into the vacated

---

2 The generalization to multiple argument functors is straightforward.

space, binding the address of the root to an identifier. The checkpointed heap fragment is now accessible and available for use. Of course, this implementation strategy is rather crude, but it is considerably easier to implement in an existing system than full object persistence, and is much more flexible than simple-minded checkpointing.

A similar approach to program persistence is used in Poly [MATT85B]. The idea is that one often does not wish to load a program into an existing context, but rather wishes to switch to the context of that program. Rather than shift the existing heap to make room for the imported heap, the current heap is merely replaced by the imported heap. This approach has the advantage over simple-minded checkpointing that it only stores the portion of the heap that is accessible from a given program, rather than storing the entire session context. However, it has the disadvantage that it destroys the current context. Although this is clearly a limitation, it appears that it is not a severe one from the point of view of program persistence.

Program persistence, however it is implemented, is supported by the introduction of two primitives, import and export. The effect of export S to H is to write the fragment of the heap rooted at structure S, together with its signature, to persistent storage, filed under the handle H. The exported heap fragment constitutes an address space in its own right, which we access by using the import primitive. The structure binding

```
structure S : SIG = import H
```

reloads the heap fragment filed under handle H, and binds to the identifier S. This operation succeeds only if the signature associated with the structure stored at H is no more general than SIG, which guarantees type safety.

It is important to note that sharing *between* two programs (structure hierarchies) is not preserved across calls to export and import. The reason for this is that the structure S and all of the objects on which it depends is culled from the context in which it resides, and written as a single entity. If some other structure incorporates one of the components of the structure being exported, then this connection will be lost when the structure is re-imported. For example, suppose that S is the root structure of some program that incorporates structure R as a substructure, and that some other structure T, unrelated to S also incorporates R. Then after exporting S to persistent storage, and re-importing it at some later time, the relationship between T and S no longer holds. In fact, T may not even exist in the context into which S is imported. From the point of view of supporting program persistence, this is perfectly alright, since only the sharing *within* a program is relevant. Sharing between programs is purely an accident of no particular importance.

In Section 2 we discussed the role of functors as providing the means of relinking programs, and therefore lie at the heart of the dynamics of program development. The key idea is that the dependency between structures can be expressed functionally, provided that we have sharing clauses. A change to a program component is made by changing the corresponding functor, recompiling that functor (by rebinding it), and then relinking the entire program. Thus functors are the "object modules" of ML, and functor application performs linking. Separately-compiled program units and program libraries are naturally viewed as persistent (pure) functors.

Since a functor is an object, object persistence provides the means of supporting separately compiled program units and program libraries. But since functors have no non-local references to the heap, they may be isolated from the rest of the heap, much as we isolated structure hierarchies above. As a result, a simpler implementation strategy is available for the support of persistent functors based on the import and export primitives defined above. Here the non-destructive form of import must be used, for otherwise there would be no structures available to apply it to once we had retrieved it!

We extend the import and export primitives to functors as follows. The command export F to H writes the functor F to persistent storage at handle H. The parameter and result signatures and the parameter sharing clause are stored with the functor at H. A stored functor is re-imported with the following variant of a functor binding:

```
functor G( R : SIGR ) : SIG = import H
```

This binding succeeds only if the signature stored with the functor at H matches the given signature of G in the following sense. Suppose that the functor stored at H has the signature <SIGR',SIG'>. Then the declaration of G succeeds only if

1) Signature SIGR is no more general than SIGR';
2) Signature SIG is at least as general as SIG'.

Note carefully the inversion between the two conditions. Recall that we say that a signature SIG is more general than SIG' if every structure that matches SIG' also matches SIG. An argument to the functor G must match the signature SIGR. If SIGR is no more general than SIGR', then any argument to G matches SIGR' as well, and hence is acceptable as an argument to the stored functor. Conversely, the stored functor yields a structure that matches signature SIG'. This structure also matches SIG if SIG is at least as general as SIG'.

It is interesting to compare our approach to persistence with the persistent abstractions of Cardelli and MacQueen [CM85]. Their interest is primarily with persistent abstract data types, and the problem of ensuring type safety in a persistent environment for a language with user-defined abstract types. Signatures are similar to their existential types, and structures are similar to their implementations of an existential type. However, we do not have any notion of a value with abstract type (which is an object that is "really" an object of the implementation type, but is treated as having a type distinct from the hidden implementation), and therefore most of the complications do not arise. They have no analog of functors or functor types in their paper. They do define some primitive operations for a persistent programming language, and we can cast our operations on structures in their terms as follows. The command export S to H can be expressed in their notation as extern(dynamic(S,SIG),H, where SIG is a signature matched by S. The declaration form structure S:SIG = import H can be expressed as structure S = coerce intern(H) to SIG

4. CONCLUSION

The general notion of object persistence, adapted to the context of modular programming in Standard ML, provides an adequate framework for addressing some of the pragmatic aspects of program development such as session preservation, separate compilation, and program libraries. Three particularly useful forms of persistence, session persistence, structure persistence, and pure functor persistence, address certain special problems that arise in a program development environment, and admit simpler implementations than a persistent heap. Whether or not full persistence is necessary for our purposes remains unclear, though it appears that the real advantages of persistence lie in the context of database programming.

5. ACKNOWLEDGEMENTS

Kevin Mitchell and Dave MacQueen carefully read a draft of this paper and made many valuable suggestions.

REFERENCES

[ACC81] [AGO85] [AM85b] [CM85] [HARP86] [MACQ85] [MATT85B] [MILN85]

Chapter 3

Persistence and Type Abstraction

Luca Cardelli
AT&T Bell Laboratories

David MacQueen
AT&T Bell Laboratories

ABSTRACT *Abstract types are a familiar and effective way of structuring programs. The basic idea of information hiding must, however, be reconciled with the need to store data for long periods of time and make it accessible to different activities. In particular, this requires that a type checker be able to recognize occurrences of the same abstract type during different activations of a program, while at the same time enforcing the privacy of data representations.*

To support such a type checker, the persistent storage of data must preserve type information, and must respect type abstraction. The use of type abstractions in the presence of persistent storage requires that abstract types be made persistent as well. Under these conditions, we can preserve type security across distinct activations of the typechecker.

The following is a brief account of how various models of abstraction and persistence interact. We start by sketching a simple polymorphic language and its types and showing various ways of modeling type abstraction in such a language. We then discuss some basic notions underlying persistent storage of typed objects, such as the intern and extern primitives and the special type dynamic, and describe three persistence strategies. Finally we discuss a particular approach to persistent abstract types.

1. VALUES AND TYPES

We will base our discussion on a simple polymorphic language in the tradition of ML [MILN84] and Amber [CARD85]. The simplified language we have in mind is closely related to the language SOL [MP85], variants of which are described in [REYN85] and [CW85].

The basis of this language is a slightly sugared applied lambda calculus that is adequate for expressing certain kinds of values. Type annotations are added to the basic expressions in such a way that one can statically determine a type for each expression. This type is a structural characterization of the value denoted by the expression. Types are viewed as meta-level terms in a type language designed to express certain structural properties of values. Types are not values, but can be interpreted semantically as sets of values.

A value may have many types, and correspondingly a value expression can be typed in several ways. For example, the "type-free" expression λx. x is the basis for the following type-annotated versions, each of which has a different type:

$$
\begin{array}{ll}
\lambda x{:}int.\ x & : int \to int \\
\lambda x{:}bool.\ x & : bool \to bool \\
\forall t.\ \lambda x{:}t.\ x & : \forall t.\ t \to t
\end{array}
$$

These type annotations do not affect the value computed by the expression (the identity function in this case), they only help to characterize the structure of that value. The last type is a polymorphic type, expressing the fact that the underlying expression λx. x can have many types, namely all instances of the type schema t → t.

1.1. Language of Types

The class of type expressions is defined by the following abstract syntax:

$$
\sigma ::= \\
int \mid bool \mid \dots \mid t \mid \sigma \times \sigma \mid \sigma + \sigma \mid \sigma \to \sigma \mid \forall t.\ \sigma\ (t)
$$

Types are *structural*, *i.e.* types are equivalent if their term structure matches (modulo change of bound type variables). Certain types are *atomic* in that they have no internal structure and match only themselves. The atomic types include certain primitive types such as int and bool, and also the *abstract* types discussed below.

A closed value expression (*i.e.* one that contains no free occurrences of unbound or lambda-bound variables and therefore denotes a particular value) must have a type that is a closed type expression, *i.e.* its type may not contain free type variables.

1.2. Existential Types and Packages

Following SOL [MP85] we will introduce type abstraction through the notion of existentially quantified types. We introduce a new class of existentially quantified type expressions:

$$
\sigma ::= \exists t.\ \sigma(t)
$$

Roughly speaking, a value has type ∃ t. σ(t) if it has type σ(τ) for some particular type τ. An expression having existential type ∃ t. σ(t) must specify a particular type τ as a witness for the existential type quantifier, and an expression having the type σ(τ). We use the following syntax:

$$
\textbf{pack}\ [\ t{=}\tau;\ e{:}\ \sigma(t)\ ]\ :\ \exists t.\ \sigma(t) \tag{1}
$$

Such expressions (and their values) will be called *packages*. In order to establish the typing (1) it is necessary and sufficient to establish

$$
e : \sigma(\tau)
$$

The type τ is called the *representation type* and the expression e defines the *interpretation* of the type t. The expression e typically denotes a tuple of values and functions through which we are allowed to create and manipulate values of the representation type. The existential type $\exists$ t. σ(t) is called the *interface* or *signature* of the package, which in turn is called an *implementation* of the interface. Matching of existential types is also structural (modulo renaming of bound variables).

For example here is a package point with representation int $\times$ int and interpretation (*i.e.* operations) origin and move (we assume for convenience that our language admits labeled product types and the corresponding labeled tuple expressions):

> **signature** Point = $\exists$ t. <origin: t, move: (t $\times$ int $\times$ int) $\rightarrow$ t>
> **value** point =
>   **pack**
>   [t = int $\times$ int;
>   <origin=<0,0>, move=**fun**(p:t,dx:int,dy:int). <fst(p)+dx,snd(p)+dy>>
>   : <origin: t, move: (t $\times$ int $\times$ int) $\rightarrow$ t>]
>   : Point

The only way to make use of a package is to *open* it in a limited scope consisting of an expression e:

> **open** A **as** t,v **in** e   : τ

where A: $\exists$ s. σ(s) and e: τ under the assumption v: σ(τ). Here t is a bound type variable, and v is a bound variable pattern of type σ(t). For example, we can use the point package to generate the point (2,3) in the following way:

> **open** Point **as** t, <origin=p, move=f> **in** f(p,2,3)

Here t is not used, but in general it could appear in function definitions in the body of open.

The treatment of the binding of t will differ according to the model of abstraction we adopt, as explained in the next section.

2. PACKAGES AND ABSTRACTION

There are several alternative ways of treating the type component of a package, and these alternative treatments lead to different styles of type abstraction [MACQ86].

2.1. The Transparent Witness Model

One approach, adopted for modules in ML [MACQ85], is to view the representation type simply as an accessible component of the package. For instance, if P = **pack** [t = int $\times$ int; ...] then within (**open** P **as** t, v **in** ...), t would be equivalent to int $\times$ int. In other words, the type component of a package is not at all hidden or "abstract". In this model, type abstraction is achieved by lambda abstraction with respect to a formal (and therefore necessarily *abstract*) package variable.

2.2. The Hypothetical Witness Model

A second approach is that of SOL, also adopted in [REYN85] and [CW85]. In SOL, the type component of a package (called a *data algebra*) is treated as purely "hypothetical", despite the fact that it is quite explicitly defined in the package expression.

In this model, (**open** P **as** t, v **in** ...) declares the names t and v to represent the (hypothetical) type component and interpretation of the package P during evaluation of the body expression. Within this scope, the witness variable t is treated as a new atomic type, and this type is not allowed to appear in the type of the open expression itself since the binding of t has no significance outside of it. (This restriction is violated by the previous example of an open expression).

This means that the type component of the package has no meaningful permanent identity; it can never be related to any other type except within the local scope of an open expression. Thus we refer to the type component as being hypothetical, indicating an even stronger constraint on its use than that implied by the conventional meaning of type abstraction, where the type retains its identity even though its structure is hidden.

One advantage of this very restrictive approach is that an existential type is just an ordinary type, and correspondingly, packages are just ordinary values that can be manipulated in all the usual ways, such as being defined by conditional expressions and serving as arguments and results of ordinary functions.

2.3. The Abstract Witness Model

The third approach is a compromise between the previous two and we could characterize it by saying that the type component of the package is real but "abstract". Given P = **pack** $[t = \tau; ...]$ we can refer to the witness type of P as P.t, but it is treated as an atomic type unique to the package P, and not as an abbreviation for the representation type τ. Under this interpretation, we will refer to a package as an *abstraction*, and to its type component as an abstract type.

Abstract types can appear (as atomic elements) in other type expressions, including the type of the body of an open expression. However, if we wish to continue to view packages as values and existential types as ordinary types in this model, the distinction between types and values becomes blurred and we have to impose some rather *ad hoc* constraints to preserve decidable type checking. For instance, if A,B: ∃ t. $\sigma(t)$ and we define:

C = **if** b **then** A **else** B

then we will probably require that the witness type of C does not match either the witness of A or of B.

The problem of persistent abstractions in this model is more interesting, because the use of one abstraction in defining another can give rise to dependencies that need to be preserved when values and abstractions are made persistent.

3. INTERN AND EXTERN

A value in main memory can be made persistent [ABCCM83] by an (explicit or implicit) extern operation which writes it to persistent memory. Such a value can then be recovered by a symmetrical (explicit or implicit) intern operation on it. Extern and intern preserve the structure of data, including sharing and circularities. Given a value A, intern(extern(A)) should return a value B which is at least isomorphic, modulo relocation, to A, and (depending on the implementation of intern-extern) can be A itself.

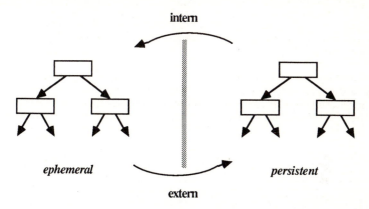

The precise way in which external values and internal values are related can change, according to the persistence model one adopts; a few alternatives will be sketched later. Independently of these different models, the basic idea is to be able to transfer arbitrary objects from main to persistent storage and back.

We assume we can intern-extern any value in memory, including programs and types. This allows us to make both types and values persistent. This ability to intern-extern values and types uniformly is useful in situations where values and types are mixed, as is the case for dynamic values, described in the next section.

Intern and extern can be implemented in terms of a few basic storage formats, like byte strings and pointer arrays. Values and types can then be represented using the same storage formats, so that intern-extern will work uniformly on both values and types.

4. DYNAMICS

What is the type of an object returned by intern? This may be difficult to determine statically, as intern can read objects of any kind. Intern and extern primitives can be safely embedded in a strongly typed system by the use of *dynamic* objects [CARD85]. The intuition is that an object of type dynamic is dynamically typechecked in the context of an otherwise statically typed language.

An object of type dynamic is a pair consisting of a type and an object of that type. Hence a dynamic object is self-describing, and can be manipulated, stored and retrieved without the usual restrictions imposed by static typechecking. A dynamic object is created by the syntax:

dynamic(τ,e)   : dynamic

A dynamic object can be coerced to a given, statically known, type by:

coerce d **to** τ : τ

If the specified type τ matches the internal type of the dynamic, then the corresponding value is returned, stripped of the type. Otherwise a run-time type error is generated.

As dynamic objects are self-describing, they are well suited to be exported to persistent storage. They also allow us to preserve strong typing in situations where static typing is impossible, as when data has a longer life span than activations of the compiler performing typechecking.

5. IDENTIFICATIONS AND HANDLES

Identifications provide an unambiguous way of telling whether two objects are the *same* object. Identifications are made unique across all systems and users which may refer to them, usually by encoding their time and place of creation within them.

An identified object is an object permanently associated with an identification. At the time of creation of the identified object, a new identification is also created for it. The identification part of an identified object can be read and compared with other identifications, but cannot be modified.

Distinct identifications will be used (*a*) to mark external forms of abstract data types, and (*b*) to mark all objects, in those models of persistent store which rely on persistent identifiers (PIDs).

Handles are names interpreted through an external persistent environment that maps them to external forms of persistent objects. For example, if a single persistent object is stored in a file, then the file name may be used as its handle.

6. PERSISTENCE STRATEGIES

We are going to sketch here three different persistence models, which correspond to three different semantics for intern-extern. In the simplest model, intern-extern work on individual values. In a more elaborate model, they work on the database as a whole. Finally, they synchronize access to a shared database.

6.1. The Fetch-Store Model

In the first scenario, persistent memory is just backup storage for ephemeral structures. The association between internal and external objects is mediated by handles, *e.g.* explicit file names.

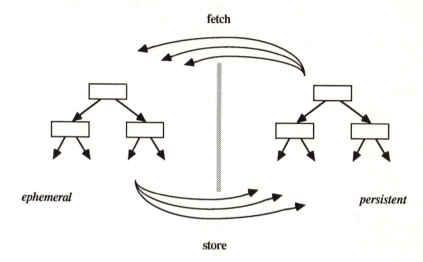

fetch

ephemeral   *persistent*

store

Extern makes a copy of an ephemeral object in persistent storage, associating it with a handle. Many calls to extern on the same object and different handles will make many independent copies. Calls of extern on two objects which share a substructure will duplicate the substructure.

Intern, given a handle, makes a copy of the persistent object in ephemeral storage. Many calls to intern on the same handle will make independent copies. Sharing is only preserved within persistent objects, not across them.

6.2. The Load-Dump Model

In the second scenario, we assume that one user has exclusive access to the persistent storage. Ephemeral memory is used as a cache for persistent storage. A one-to-one association between internal and external objects is maintained though the use of PIDs. Each object is identified by a PID, both in persistent and in ephemeral storage.

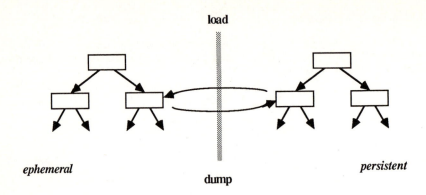

ephemeral   *persistent*

dump

Extern makes a copy of an ephemeral object in persistent storage. All calls to extern on the same object will produce the same persistent object. Calls of extern on two objects which share a sub-object will preserve the sharing of the sub-object, as the common sub-object has its own PID. Intern makes a copy of a persistent object in ephemeral storage. All calls to intern on the same PID will give the same object. Sharing of sub-objects is preserved.

If intern-extern are automatically called when needed, this achieves the semantics of loading the whole database at the beginning of execution, and dumping the whole database at the end. Intern-extern are only used to load and dump the database incrementally.

This strategy is analogous to virtual memory, where virtual addresses play the role of PIDs.

6.3. The Lock-Commit Model

In the third scenario, ephemeral memory is a cache for shared persistent storage with concurrent access. Intern-extern can no longer be fully automatic, because other users or processes may be affected.

Intern-extern work as in the previous case, through PIDs. Intern-extern must be explicit again, as in the first scenario, to control the synchronization aspects. Intern may be made to correspond to *lock* and extern to *commit*.

The load-dump model of persistence is the simplest one, conceptually, as intern-extern are automatically performed. Unfortunately it does not scale up to concurrent access (lock-commit), for which we need explicit intern-extern operations in order to perform atomic transactions. This seems to be a point in favor of fetch-store with respect to load-dump, because the former is compatible with lock-commit.

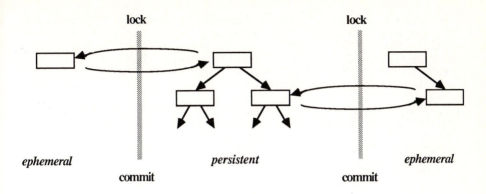

lock   lock

ephemeral   *persistent*   *ephemeral*

commit   commit

7. PERSISTENT ABSTRACTIONS

We now consider the interactions between models of abstraction and models of persistence. Under the hypothetical witness model, packages are not distinguished from other values and the witness type has no permanent identity, so the usual treatment of intern and extern for dynamic values applies to packages as well. Hence we will deal hereafter with the abstract witness model; it appears that the transparent model can be treated in a similar fashion.

The load-dump model of persistent storage behaves exactly like a single programming session, and nothing special has to be done to provide persistent abstractions or values. Use of dynamics and the need for persistently identifying abstract types arises primarily in the fetch-store model, and in the more general lock-commit model. Hence we shall concentrate here on the latter models.

7.1. Abstract Witnesses in the Fetch/Store Model

An abstraction consists of:

> (1) an identified object representing the abstract type,
> (2) a tuple of operations (the interpretation), and
> (3) the interface specification.

Optionally the representation type itself may be included for debugging purposes.

(1) and (2) constitute the abstraction proper, which implements the interface (3). The interface is expressed as an existential type, *i.e.* it is a structural type not involving the abstract type. These components are all that is needed for manipulating objects and typechecking expressions of that type.

There is an analogy between an abstraction and its interface and dynamic objects. A dynamic object can be considered as having the interface: $\exists t.\ t$. However the type component of a dynamic object is not abstract, since it can be inspected by the coerce statement and matched against the specified type.

An extern operation on an abstraction moves all the above pieces of information to persistent storage. One should make sure that an abstraction is made persistent before making objects of its type persistent. To guarantee this, abstractions might automatically become persistent at the time of their creation, which would require that a handle be automatically generated for each one. Alternatively this could be accomplished by explicitly externing the abstraction.

When externing an object of an abstract type we create a dependency between (*a*) the object and (*b*) the abstraction which carries the abstract type and the operations which are supposed to work on that object. When we intern such an object, the corresponding abstraction must also be fetched, to provide a context supporting the use of the abstract object.

For example, we could at first define a new abstraction point and extern it together with a value of its type:

```
signature Point =...
abstraction point = pack ...
extern("point", dynamic(Point,point))
open point as t,<origin=o,move=m>
in extern("p",dynamic(t, m(o,2,3)))
```

Later on, perhaps in a different session, we intern the abstraction in order to intern the value:

```
signature Point =...
abstraction point = coerce (intern "point") to Point
value p = open A as t,p in coerce (intern "p") to t
```

As a more complex example, we can parameterize with respect to an abstraction and a handle:

```
λAbs:Sig.  λx:handle.
    open Abs as t,p in
        pack [t'=t; (coerce (intern x) to t × (t→int)) : t' × (t'→int)]

: Sig → handle → ∃t'. t' × (t'→int)
```

Here the target type of the coercion is determined dynamically, as it must be extracted from the abstraction parameter. The resulting type of the function does not depend on it, because pack abstracts it away by existential quantification. The resulting package is a new abstraction, which is independent of the original parameter abstraction and therefore must be self-sufficient.

In a slightly more general situation, we may want to extern an object that is not of a given abstract type, but whose type involves the abstract type, *e.g.* a function having abstract types as parameters and results. Such an object is necessarily associated with the abstraction because the abstract type occurs in its type, and its full use may require the presence of the abstraction. Hence the abstraction itself should be persistent and should be interned along with the object that depends on it.

8. CONCLUSIONS

We have discussed several models of persistence and abstraction. The main idea is that of preserving the identity of abstract types by giving them persistent identifications. This can be achieved in slightly different ways in the different schemes of persistence and abstraction. We have mainly dealt with the combination of the abstract witness model of packages with the load-fetch model of persistence.

References

[ABCCM83] [CARD85] [CW85] [MACQ85] [MACQ86] [MILN84] [MP85] [REYN85]

Chapter 4

An Overview of the Poly Programming Language

David C.J. Matthews
University of Cambridge

ABSTRACT *Poly is a general purpose programming language based on the idea of treating types as first-class values. It can support polymorphic operations by passing types as parameters to procedures, and abstract types and parameterised types by returning types as results.*

Although Poly is not intended specifically as a database programming language it was convenient to implement it in a persistent storage system. This allows the user to retain data structures from one session to the next and can support large programming systems such as the Poly compiler and a Standard ML system.

1. POLY AND ITS TYPE SYSTEM

Poly [MATT85a] is based on the idea of types as first class values first used in the language Russell [DD79]. In the terms used by Cardelli and MacQueen[CM85] it uses the *abstract witness* model of a type. Treating a type this way means that polymorphism, parameterised types and modules are all handled by the general concept of function application.

1.1. Types as Values

A type in Poly is a set of values, normally functions. For example the type *integer* has operations +, - etc. Other types may have these operations, the type *real* also has + and - but will not have a *mod* (remainder) operation. The operations need not be functions, *integer* also has *zero, first* and *last* which are *simple values*, and other types may contain types.

All values in Poly have a *signature*, called a *specification* in earlier reports, which is only used at compile-time. It is the analogue of a type in languages like Pascal and corresponds in many ways to the idea of a type in Ponder[FAIR82]. There are three classes of value in Poly, the *simple value* which corresponds to what are normally thought of as values in, say Pascal, numbers, strings, vectors etc.; the *procedure* or function which operates on values and the *type* which is a set of values. Each kind of value has a signature.

To show why this view of types is useful we will consider some properties of other languages, and how they are handled in Poly.

1.2. Polymorphism

A polymorphic function is one that can be applied to values of many different types. The phrase is sometimes used where *overloading* would be more appropriate, for example the + operator in Pascal. In Pascal, or languages like it, there are operators which can be applied to values of more than one data type and their meanings are different according to the type of their arguments. They can be thought of as a set of overloaded operators in the same way as operators in Ada can be overloaded. Truly polymorphic functions are somewhat different. They are functions which are applicable to values of a wide variety of data types, including types which may not exist at the time the function is written. The fundamental difference is that a new polymorphic function can be written in terms of other polymorphic functions, while a function written in terms of overloaded functions must be defined for each data type even if the program is the same for each. For example

```
function min( i,j : integer ) : integer
         begin
         if i < j then min := i else min := j
         end ;

function min( i,j : real ) : real
         begin
         if i < j then min := i else min := j
         end ;
```

The ML[MILN84] programming language provides polymorphic operations on an all-or-nothing basis. This allows one to write an identity function which simply returns its argument, and this function is applicable to values of any type. One can also write functions which operate on lists of any type or on functions of any type. This generally works very well but has problems when one wants to write an operation which operates differently on different data types. For example it is still necessary to overload = since comparing two integers is different to comparing two lists of real numbers. The *min* function cannot be written as a single function in ML. What is required is a way of writing operations which are *type-dependent*.

A type in Poly is characterised by the operations it has. Both *real* and *integer* have < operations though they will be implemented in different ways. Many other types may have < operations since Poly allows the user to make new types. Poly allows a function to be written which selects certain operations from a type and values of any type with those operations can be used as a parameter. For example there is a *single* < function which works on types which have a < operation and simply applies the operations to the arguments. The effect is as though < were being overloaded. However, we can write a function in terms of this, such as the *min* function. This will also work on values of any type which has a < operation. For example, *min* is a function which will work on values of any type with the < operation. Such a type has signature

```
type( t ) < : proc( t ; t ) boolean end
```

This type has an operation, <, which takes two values and returns a *boolean*. We will first write a version of *min* which takes three parameters; a type and two values of this type and returns a value of the type. It has signature:

```
proc( t : type( t ) < : proc( t ; t ) boolean end ; t ; t ) t
```

We can write the whole function.

```
let min ==
proc( t : type( t ) < : proc( t ; t ) boolean end ; x,y : t ) t
      begin
      if x < y then x else y
      end ;
```

It can be applied to integer values

```
min( integer, 1,2 )
```

or string values

```
min( string, "abc","abd" )
```

or values of any type with a < operation.

The first parameter is a type which must have a < operation which compares two values of the type, and the second and third parameters must be values of the type. When we call

```
min( integer, 1,2 )
```

the actual parameters are matched to the formal parameters from left to right. First the types are matched by checking that the type given has the appropriate operation, and then the values are matched. They are not of course the same type as *t*, since they have type *integer*, but we invoke a matching rule which says that if we have matched an actual type parameter to a formal type then we can match values of corresponding types. In addition the type of the result becomes matched so that the result has type *integer*. This can be thought of as a systematic renaming of *t* with *integer*.

1.3. Implied Parameters

Having to pass the types explicitly is often a nuisance so there is a sugared form which gives a way of omitting the types and having the compiler insert them automatically using the types of the parameters. The only difference to the definition of the function is that the types are written in square brackets before the other parameters. The definition of *min* would then be:

```
let min ==
proc[t : type( t ) < : proc( t ; t ) boolean end] ( x,y : t ) t
      begin
      if x < y then x else y
      end ;
```

It can be used by just giving the values.

```
min( 1,2 )
min( "abc","abd" )
```

This sugaring also allows us to define operators such as + and < which simply apply the operation with the same name from the types of their arguments giving the effect of overloading.

```
let + ==
proc infix 6[t :type(t) < :proc(t ; t)boolean end](x,y : t) t
     begin
     t$+ ( x,y )
     end ;
```

2. PARAMETERISED TYPES

So far we have seen how having types as parameters to a procedure allows us to write polymorphic operations. Types can also be returned from procedures and this provides a way of defining types which are parameterised by either types or values. As an example, suppose we wanted to construct an associative memory in which to store values of arbitrary type together with a number which would identify each. This could be defined as follows

```
let associative ==
proc( element : type end )
     type( assoc )
     enter : proc( assoc ; integer ; element ) assoc ;
     lookup : proc( assoc ; integer ) element ;
     empty : assoc
     end
     begin
          type( assoc )
          extends struct(next:assoc; index:integer; value:element);
          let empty == assoc$nil ;
          let enter ==
          proc( table : assoc ; num : integer; val : element ) assoc
               begin
               assoc$constr( table, num, val )
               end ;
          letrec lookup ==
          proc( table : assoc ; num : integer ) element
               begin
               if table = assoc$nil
               then raise not_found
               else if table.index = num
               then table.value
               else lookup( table.next, num )
               end
          end
     end ;
```

This is a very simple minded definition but it illustrates the point. We start by giving the header of the procedure which includes the signature of the argument, in this case that *element* is a type but that any type will do, and the signature of the result. The result is a type with three objects, a value which denotes the empty table and procedures to enter and look up items from the table. It is implemented in terms of a **struct** (a record with a *nil* value and equality) which makes up a list of index/value pairs. *enter* just returns a new list with the new pair "cons-ed" onto the front[1]. A better implementation would check to see if there was already an entry with that index and return a list with the old entry replaced by

---

1 We could have written simply
        let *enter* == **assoc$constr** ;
        since the arguments are in the same order.

the new one. *lookup* searches the list for an entry with the required index and either returns the value or raises an exception.

There is no particular reason why we should use integers as the indexing value, it would be perfectly possible to use any type which had an equality operation. The procedure header would then be

```
proc( element : type end ;
index_type : type( i ) = : proc( i ; i ) boolean end ) . . .
```

with *integer* replaced everywhere in the body by *index_type*.

A more efficient implementation for index types with an ordering would be to use binary trees rather than lists. We would then have to add a > or < to *index_type*, or at least replace the = by one of these. Now, since types are values we could incorporate an if-statement into the procedure and use one or other of the implementations depending on the value of a further parameter. We might want to do this because one implementation may be more efficient for, say, small tables and the other for larger ones. For the example we will assume a parameter *use_binary_tree*. The procedure will now look something like this.

```
proc( element : type end ;
index_type : type( i ) = , < : proc( i ; i ) boolean end ;
use_binary_tree : boolean ) . . .
begin
if use_binary_tree
then
    type . . . { Binary tree implementation }
      end
else
    type . . . { List implementation }
      end
end
```

This could now be called as

```
let a_table == associative( string,integer,true ) ;
let another_table == associative( string,integer,size > 30 ) ;
```

In the second case the expression may not be able to be evaluated when the call to the procedure is compiled, *but this does not matter*. We do not know at compile-time which of the two implementations of the type will be used, but we know that either of them have all the operations required so they will do equally well.

There is however a problem with this idea of types which this example shows quite nicely. Since the expression may not be evaluated at compile-time how do we know when two values have the same type? The type system must ensure that we apply the *lookup* procedure which understands the representation of the particular associative memory. It would be catastrophic to try to look up a value assuming that the value represented a tree when it was in fact a list. We need the type system to assure us at compile-time that the expressions

```
let y == X$enter( X$empty,1,"hello" ) ;
X$lookup( y ) ;
```

where ⅹ stands for a type or type-returning expression, will not give faults at run-time because of a mistake in interpreting the representations.

There are several possible approaches to the problem of which Poly and Russell illustrate two. In Russell values can have types such as

```
associative( string, integer, size > 30 )
```

provided nothing in the expression involves a global variable[2]. This essentially means that all functions have to be "variable-free", not just those which directly return types. Given this restriction it is possible to say that if two expressions are syntatically the same in a given context then they return the same value. If however, *size* were a variable, or *associative* looked at the value of a global variable, then we could not say with certainty that two values with type

```
associative( string, integer, size > 30 )
```

had the same type. Taking a purely synatactic view means that expressions like

```
associative( string, integer, 2 > 1 )
```

and

```
associative( string, integer, true )
```

are not the same type.

In Poly types are only regarded as the same if they are the same *named* type. So while values with types which are expressions can sometimes be produced there is very little that can be done with them. To be useful a type-returning expression has to be bound to an identifier.

```
let a_table == associative( string, integer, true ) ;
let a_val == a_table$enter( a_table$empty, 1, "hello" ) ;
let another_table == associative( string, integer, true ) ;
let another_val ==another_table$enter(another_table$empty, 1, "hello");
```

a_val and *another_val* have distinct types *a_table* and *another_table*.

A side-effect of this is that "types" such as

```
list( integer )
```

cannot be used directly. We have to write

```
let int_list == list( integer ) ;
```

---

2   Variable in this context means something whose value can be changed by assignment..

and then use *int_list* as the type. However this is not such a problem as might at first appear. Since we can write functions which take implied parameters we can write an *append* function which will work on values of any type with the appropriate *hd*, *tl* etc., irrespective of their actual implementations.

3. MODULES

A module is conventionally thought of as a collection of types and functions which can be separately compiled. It has an interface which is the types of these functions so that other modules can make use of it without having to know the precise implementation.

Types in Poly can be thought of in the same way. A type is a collection of operations and its signature gives their "types"[3].. A module which makes use of other modules, *imports* them in conventional terms, can be represented as a procedure which is applied to types and returns a type. One of the big advantages of this view of modules is that binding modules together is done using statements written in Poly and type-checked using the normal Poly type-checker. There is no need, as with MESA and C-MESA[MITC79] for a separate module binding language.

The module system for ML[HARP85] is essentially a system built on top of the kernel language. *Structures* and *functors* correspond to values and functions in the kernel but the ML type system makes it impossible to unify these concepts.

4. PERSISTENCE IN POLY

Poly is an interactive system in which the user types expressions and declarations and these are compiled and executed immediately. When objects are declared they are added to the objects the system knows about and they can be used in subsequent expressions. Such systems are quite common and usually work on a core image which can be saved from one session to the next. This is fine provided that the core image does not grow too big. However as the core image gets larger the costs of reading it in and writing it out get more serious. Also the cost of garbage-collection rises. There is a further question about the security of the data if the machine crashes while writing out a large image.

For these reasons Poly is implemented in a persistent store [ACC81] [ABCCM81] which can be thought of as a core image where objects are only read in when they are actually required. The cost of loading objects from the image, or database, depends on the amount of the store which is used by a program rather than the total size of the image. A simple transaction mechanism ensures that the database remains in a consistent state in the event of a machine crash or if the program is killed halfway through writing out. Some experiments have been done on using multiple databases so that large programs such as the compiler can be shared between several users.

---

3   We usually think of a type as being something like integer which has values, but a type in Poly can be any collection of objects. So a collection of floating point functions sin cos etc. could be combined as a type even though there is no such thing as a value of this type.

Using this persistent store the Poly compiler has been boot-strapped so that it is just another procedure. A Standard ML compiler has also been written which uses the same back-end as the Poly compiler.

In a typical interactive programming system there is a single name space for all identifiers, but as the number of declarations have grown it has become necessary to divide up the name space into separate *environments*. An environment is very similar to a directory in a filing system or to a block in a programming language. When an environment is selected all new identifiers are entered into it and looked up in it. There is the equivalent of the scope rules in a programming language so that an identifier is looked up in a series of nested environments until it is found. It could be thought of as a Poly type since it is a collection of objects, but it cannot be quite the same because declarations can be added or removed dynamically to an environment while a Poly type must be "frozen".

5. CONCLUSIONS

Poly was designed as a general purpose language and has been used successfully for some medium scale projects (there is about 20000 lines of code in the Poly and ML compilers). After some years of programming in it the type system has proved to work very well. Treating types as first-class values seems to result in a generally simpler language than languages where types are treated as purely compile-time objects. Experience with Standard ML suggests that pattern-matching and exceptions with parameters (exceptions in Poly cannot carry parameters) are something that should be added. Some kind of type inference based on unification could be used to reduce the amount of type information which must be given explicitly, though it cannot remove it completely.

The presence of a persistent store tends to break down the distinction between compile-time and run-time, since the compiler is just another function to be applied. Compile-time does have some meaning in this system however. Compiling an expression means checking the interfaces between functions and their arguments so that the result can be guaranteed not to produce a type-checking error later on. If we compile a procedure then we want to produce a type for the procedure as a whole and remove the type information within it. Not only does this improve the efficiency of the procedure but it also gives us a degree of certainty that the procedure will not fail. It is a little way along the road to proving the correctness of the procedure. There is a cost in this static type checking in Poly in that some procedures which are in fact type-correct will fail to pass a static type-checker, but the advantages of static type-checking more than outweigh the disadvantages.

REFERENCES

[ACC81] [ABCCM81] [CM85] [DD79] [FAIR85] [HARP85] [MATT85] [MILN84] [MITC79]

Chapter 5

Functional Databases, Functional Languages

Rishiyur S. Nikhil[1]
Massachusetts Institute of Technology

ABSTRACT *Database systems today have evolved a great deal from the first storage structures, towards greater data independence, expressive power in manipulation languages, and expressive power in data models. But their facilities are still poor substitutes for analogues in programming languages: data abstraction, structured control constructs and type systems respectively. Most programming languages, on the other hand, deal inadequately (if at all) with the question of long-lived structured data. The problem is compounded when dealing with both a database system and a programming language that are alien to each other (as is common today).*

Functional programming languages with functional databases offer a clean solution to this problem. Functional programming languages are expressive, do not have destructive updates, and allow much parallelism. Functional databases are never destructively updated—rather, one views them as infinite sequences of versions.

After outlining the features of such a system, we discuss its many advantages pertaining to types, optimization, expressive power, and interactive environments.

1. WHAT IS WRONG WITH DATABASES TODAY?

Today's database systems evolved out of mass-storage device architectures. This evolution has moved away steadily from device-specific models to more and more abstract models— for example, relational databases. Still, they are so difficult to use that many applications do so only when efficient and reliable access to large amounts of data is of over-riding concern. The designers of hundreds of everyday applications (mail systems, software engineering aids, word processors, ...) prefer *not* to deal with the complexities of current database systems and use primitive file systems instead.

Let us examine some of the characteristics that make current database systems so difficult and inconvenient to use.

---

[1] This research was done at the MIT Laboratory for Computer Science. Funding for this project was provided in part by the Advanced Research Projects Agency of the Department of Defense under the Office of Naval Research contract N00014-84-K-1025.

1.1. Inadequate Type Structure

To a first approximation, the term Data Model used in databases is analogous to the term Type System in programming languages, *i.e.*, it specifies a universe of definable types. A schema for a database is analogous to a particular type definition within that type system, and a database itself is analogous to a particular value (data structure) of that type.

The type systems of most database systems are extremely poor in comparison to those available in modern programming languages. For example, CODASYL databases [DATE86] provide scalar types, tuples (*i.e.*, records) of scalars, sequences of records and record-sequence pairs (owner and members of a "set"). In addition, there are various means to share records amongst different sets. Relational databases provide scalar types, tuples of scalars, and sets of tuples (*i.e.*, relations).

Unfortunately, these structuring mechanisms are not suitable for many applications - the encoding of data from the problem domain into (say) relations can be extremely awkward, and is, thus, error-prone. For example, circuit diagrams, graphs, IC layouts, matrices, *etc.* while not impossible to represent relationally[2], stretch the model so much as to lose the progammer's intuition. Forcing the use of *only* relations greatly increases the programmer's intellectual burden, and is not unlike trying to encode complicated structures such as graphs and trees into FORTRAN arrays.

Some of the newer so-called Semantic Data Models attempt to remedy this shortcoming [BMS84]. But they are still poor substitutes for the type facilities of modern programming languages. Many of them permit the description only of passive collections of data— they do not integrate data-manipulation *procedures* as part of the schema for a particular database. Instead, there is typically only a single, pre-determined generic set of database operators. This separation leads to two problems. First, it does not allow data abstraction, about which we will say more in the next section. Second, it makes it difficult to mix intensionally- and extensionally-defined data smoothly.

A serious shortcoming of most proposed data models is that they are rarely accompanied by rigorous semantics. Even in the relational model (one of the cleanest) there are many unclear areas. For example, consider joining a PERSON-AGE-FATHER relation with itself to get a grandfather relation. To avoid producing a meaningless relation with two columns with the same heading AGE, one needs special constructs such as *renaming*, whose semantics are often unclear.

1.2. Lack of Data Abstraction

The lack of data abstraction facilities is another serious problem in most data models. In databases one of the goals has long been to achieve "data independence" [DATE86], *i.e.*, the independence of application programs from the particular access methods and storage structures used to implement a conceptual schema. Current database systems achieve this to varying degrees.

This concept is well established in programming languages. It is in fact no more than the "representation independence" of abstract data types— it is exactly what makes these types abstract. It is well known how to achieve this— each abstract type is accompanied by a set of procedures to manipulate objects of that type, and there are linguistic mechanisms to ensure those procedures are the *only* way to manipulate such objects. If the representation of an abstract type is changed, one need only change the

---

[2] See, for example, [Merrett85].

implementations of the associated procedures; all client applications will be unaware of and unaffected by the change.

It is often necessary to maintain a notion of consistency in a database beyond what it is expressible in the type system (for example, that the standard deviation of machine-part tolerances should not exceed a certain number). Updates that result in inconsistent states of the database should not be permitted. Definitions of consistency are usually application-specific and may be arbitrarily complex. Many data models rely on a separate language of integrity constraints to express consistency requirements. There are several problems with this approach. First, it is yet another language for the programmer. Second, because the constraint language is separate from the data manipulation language, it may be difficult to check constraints efficiently (*e.g.*, during the update). Third, the programmer is not left with much flexibility in error-handling— *e.g.*, to trap violations, identify reasons, and take corrective action. Thus, languages for integrity constraints are often deliberately simplistic, incapable of expressing complex consistency requirements.

In programming languages, one again uses abstract data types to maintain integrity constraints (also known as *invariants*). Abstract data types ensure that data of a particular type can be manipulated only by authorized procedures. These procedures are specified by the programmer who has the full power of the programming language available to express integrity checks as complicated as the application demands, with whatever efficiency is possible, and with appropriate error-handling. The programmer may choose to maintain the standard deviation of machine-part tolerances internal to the machine-part abstract type, recomputing it incrementally on each update. This is much more efficient than recomputing it after each update.

Alerters are another useful feature of databases. Like integrity constraints, an alerter monitors the database for some condition, except that these are not necessarily violations of consistency - *e.g.*, large monetary withdrawal, fuel-level low, *etc.* On such conditions, an alerter performs an associated action, such as sending a warning message to a responsible person. Again, it is much better to incorporate alerters into the procedure definitions in an abstract data type rather than use a separate alerter language and a separate implementation technique.

1.3. Lack of Integration with Programming Languages

Most databases can be queried and updated in two ways: using a stand-alone query/update language (*e.g.*, SQL, QUEL), or by embedding a database sublanguage in an existing programming language (*e.g.*, EQUEL in C, SQL in PL/1) [DATE86].

Stand-alone database languages are generally very convenient, but are typically limited in expressive power, lacking general control structures, procedural abstraction, full arithmetic, *etc.*

There is usually a semantic mismatch between database sublanguages and the programming languages they are embedded in, making them notoriously difficult to use. Typically, the type structures and the control structures in the programming language are quite different from those of the database sublanguage. For example, the database may be relational, but there may be no types or iteration constructs in the programming language that can naturally model relations and relational operations.

The methods to circumvent this mismatch are usually cumbersome and unsafe. For example a procedure may send a message to be interpreted by the database system, which leaves a result in some invisible work area. To examine this result, it may be necessary to establish a "cursor" on it, which must then be moved in small sequential increments revealing a small piece of the result at a time. Further, this kind of interface may relinquish any type-checking normally provided in programming language. The imposition of unnatural control structures to traverse database objects can negate any data-independence that the data model originally provides. The programmer must now manage two dissimilar models of store— that of the programming language, and that of the database system.

2. WHAT DATABASE FEATURES DO PROGRAMMING LANGUAGES LACK?

We have claimed that many database concepts have better analogues in programming languages. What then are the particular characteristics of databases that programming languages lack?

1. Databases are not *ephemeral*, *i.e.*, they are "permanent" data-structures. In contrast, most data-structures in programs have a lifetime no longer than that of the program itself.

2. Databases provide efficient access to data structures implemented in a memory heirarchy which encompasses devices with widely varying characteristics, such as latency, bandwidth, unit of data transfer, *etc.*

3. Databases are resilient. Data is remembered in spite of hardware failures.

4. Databases provide concurrency control, *i.e.*, controlled access from multiple, independent processes, such that database consistency is maintained.

5. Databases provide facilities for access control, security, *etc.*

Of course, not all databases today provide all these features! Database system designers have always attached great importance to these characteristics while slowly evolving databases away from device-specific architectures and incorporating more and more features normally found in programming languages.

3. DATABASE PROGRAMMING LANGUAGES

We believe that the bottom-up evolution of database systems outlined in the previous section can limit our view of what databases are and what they should be. We have often experienced the following situation: When we point out the poverty of today's data models in comparison with type systems in programming-language, database researchers often respond that such richness and generality is "not necessary in database work". Unfortunately, this is merely a self-fulfilling assessment— the requirements of database work tend to get defined in terms of what is currently available and feasible. We firmly believe that there is ample need for richer type systems in databases.

We are therefore exploring the opposite evolutionary path: begin with an expressive programming language, and gradually incorporate features normally found only in databases, putting to good use all the lessons learned in implementing today's database systems. Of course, this book is evidence that several researchers share this view. Let us now look at some the features we would like in an integrated system.

A primary requirement for a database programming language is a rich type system in which one can model structures in a wide variety of application domains naturally. Arbitrary objects should be storable in the database, *i.e.*, there should be no distinction between programming language types and database types. There should be user-defined types and abstract data types, with rigorous semantics. It is desirable to support subtyping and inheritance— experience with Simula and Smalltalk have shown this to be a powerful modeling tool.

Another major requirement is for the programming language itself to be a very high-level language. The higher the level, the easier it is to produce and to maintain programs. This need is especially acute in the database environment, where, almost by definition we are dealing with a long-lived, shared, continuously evolving system. Among necessary advanced features we certainly include the capability of having functions as first-class values.

There have been some attempts to integrate programming languages with databases. For example, Pascal/R [SCHM77] is a system that integrates a relational database system with Pascal. It extends the type system of Pascal to include a **relation** type built on Pascal records, and a **database** type built on relations, and it extends the control structures of Pascal to include constructs to traverse and manipulate relations. This is a very successful and elegant integration of a conventional database and programming language. This research has since evolved into the design of DBPL, a relational programming language based on Modula-2.

There are also many exciting efforts currently under way to integrate logic programming languages (such as Prolog) with relational database systems to create so-called knowledge bases [GMN84]. The concept exploited here is that ground facts in a logic program have a natural interpretation as a relation in a relational database. This approach thus gives relational databases a reasoning capability, and allows smooth integration of intensionally- and extensionally-defined data.

However, a limitation of these approaches is that the languages used are not expressive enough in their own right (in particular, no functional values). In addition, not all types may be persistent, and those that can are too limited (just flat relations, in both cases).

PS-Algol [ABCCM81] is a language in which objects of arbitrary type may persist. A database exists as a persistent heap external to the program, and there are mechanisms to make the database appear transparently as part of the standard program heap. Objects in the heap may be modified with the usual assignment statement, and any objects on the heap may persist beyond the lifetime of the program if they are made reachable from the top-level table of a database. Pointers to records are dynamically typed, giving a measure of polymorphism, and thus, type-checking is not completely static, though it is safe. PS-Algol has no abstract data type facility *per se*, although given the interesting combination of block structure, assignability of procedure values, the polymorphic procedure value **nullproc** and the fact that pointer types do not have to be declared, one can simulate abstract data types [ABCCM84]. This work has since evolved into the design of Napier, a persistent language with richer types and stronger type-checking [Atkinson85a].

Galileo [ACO85] [AGOP85] is a database programming language with a rich polymorphic type system, based on ML [CARD83b]. In Galileo, the database is part of a global environment inside which each transaction runs. The Galileo type system includes mutable objects. A database program is an expression evaluated in this environment, and this evaluation may modify the updatable objects in the database.

In both PS-Algol and Galileo, mutable objects are an integral part of the language, and database updates are expressed using the standard assignment statement that destructively modifies an object in the database. Complex updates (*e.g.*, adding 5% interest to every account with more than $1000) are performed by embedding assignments in standard control structures.

In contrast, we are exploring a database programming language with an essential difference: the database is *immutable*, for which reason we call it a *"Functional Database"*. An update transaction specifies a new version of the database in terms of the old. The same functional language is used both for queries and updates— a language in which there are no destructive assignments, and in which functions are first-class values.

We call this class of database programming languages FDBPLs, or Functional Database Programming Languages. We explicate and justify this approach in the next section, but first we summarize our view of the evolution of DBPLs below:

Conventional Programming — COBOL, PL/1, *etc.*
Language + File Structures

⇓

DB with limited types + alien PLs — Most current DBMSs

⇓

Integrated DBPL with limited types — Pascal/R,DBPL,Prolog+RDB

⇓

Integrated DBPL with rich types — PS-algol, Napier, Galileo,
FDBPL

4. A FUNCTIONAL DATABASE PROGRAMMING LANGUAGE

In this section we outline a functional database programming language (FDBPL) and discuss its advantages.[3] The syntax used in these examples is still under development and is based on the dataflow language Id Nouveau [NIKH87] [ANP87a].

4.1. What is a Database in FDBPL ?

In FDBPL a database is an *environment* of bindings — *i.e.*, a mapping from identifiers to types and values. The values named in an environment may be of any type, such as scalars, abstract types, data structures (*e.g.*, arrays and lists), and functions, but functions are central to the model.

---

[3] There seem to be some differences among various researchers in the use of the adjectives "functional" and "applicative" to describe a language. We use the terms interchangeably, with the meaning that it supports functions as values and is free from side-effects.

Common database structures, especially indexing structures, are found as subtypes of the ordinary function type (written `t1 -> t2`). These subtypes are called *database function types*.[4]

To motivate this uniform functional view, let us draw an analogy with arrays. Abstractly, an array A of objects of type *t* is just a function from integers to *t* that is defined on a finite sub-domain of integers. To index the array is to apply the function to an integer argument. However, because it is a specialization of ordinary functions, an array admits additional operations, such as enumeration of the sub-domain on which it is defined. In FDBPL we generalize this idea to encompass common database structures:

- The domain of a database function can be any type.
- Database functions often have associated inverses.
- The range of a database function is often a list.

To capture these variations, we introduce several database function type constructors which are specializations of the general "->" type constructor:

```
t1    -->   t2                          (1)
t1    -->*  t2                          (2)
t1    <-->  t2                          (3)
t1    <-->* t2                          (4)
t1    *<-->* t2                         (5)
```

A function `f` with any of these types has domain `t1` and range either `t2` (if the forward arrow is `>`), or `*t2`, *i.e.*, lists of `t2` (if the forward arrow is `>*`). In cases (3), (4) and (5), it also has an inverse named `^f` with domain `t2` and range `t1` or `*t2` depending on the reverse arrow.

For example, here is the type of an environment (*i.e.*, the schema) for a student-course database:

```
Student      : TYPE
mkStudent    : Void -> Student
Students     : *Student
SName        : Student   <-->   String
SStatus      : Student    -->   String
STotalUnits  : Student     ->   Number

Course       : TYPE
mkCourse     : Void -> Course
Courses      : *Course
CName        : Course   <-->   String
CUnits       : Course    -->   Number
CPrereq      : Course *<-->* Course

Enrollment   : TYPE
mkEnrollment : Void -> Enrollment
Enrollments  : *Enrollment
EGrade       : Enrollment --> String

S-enroll     : Student <-->* Enrollment
C-enroll     : Course <-->* Enrollment
```

---

[4] Ordinary functions can be regarded as intensional, or programmed functions. Database functions can be regarded as extensional, or indexed functions.

Here Student, Course and Enrollment are database types. The name mkStudent is bound to a function of no arguments that creates a new Student object. The name Students is bound to a list of students (the "*" is a unary prefix type-constructor meaning "list of"). SName is a 1-1 function between Students and Strings; this means that SName applied to a Student returns a String and ^SName applied to a String returns the Student with that name. SStatus maps Students to Strings. STotalUnits is an ordinary function from Students to Numbers.

CPrereq is a many-to-many function between Courses and Courses. When applied to a Course, it returns a list (perhaps empty) of Courses that are its prerequisites. When ^CPrereq is applied to a Course, it returns a list of Courses for which it is a prerequisite.

S-enroll is a 1-to-many mapping between Students and Enrollments; when S-enroll is applied to a Student it returns a list of his/her Enrollments, and ^S-enroll applied to an Enrollment returns the Student corresponding to that Enrollment.

In conventional databases, Student, Course and Enrollment would be record types (tuple types), Students, Courses and Enrollments would be relations, SName and CName would be key fields, SStatus, CUnits would be ordinary fields, and functions like CPrereq and S-enroll and C-enroll would correspond to various joins between relations.

4.1.1. Queries

A *Query* is merely a functional expression evaluated in the database environment, producing its value as the answer. In functional languages, the treatment of functions as values is central, allowing the use of poweful, bulk operators to express computations concisely [TURN81]. Examples of such operators are:

- (map f l) returns a list containing the results of applying the function f to each member of the list l.

- (filter p l) returns a list containing just those members of l that satisfy the predicate p.

- (fold op v l) returns an accumulated value over the list l, obtained by applying the binary function op pairwise to each element of l, with initial value v. For example, fold (+) 0 l sums the list l.

Functions are curried, so that (fold (+) 0) is a function that takes a list of numbers as its argument and sums it. The advantage of treating all database structures as functions is uniformity — one can then extend all the power of these high-level operators to database structures.

For example, here is a query to find the names of all special-status students taking 15-unit courses:

```
{  e_has_15 e            = (CUnits (^c-enroll e) == 15) ;
   es_with_15            = filter e_has_15 Enrollments ;
   ss_with_es_with_15    = map ^s-enroll es_with_15 ;
   special_s s           = (SStatus s) == "special"
 In
   map SName
       (filter special_s
               ss_with_es_with_15) }
```

We define a temporary predicate e_has_15 that decides if the course related to an Enrollment object has 15 units. Using it, we filter all enrollments to compute es_with_15, the list of all 15-unit enrollments. Using ^s-enroll, we map it into a list of the corresponding students. Finally, we filter it with the predicate special_s to get the students of interest, and map SName over that list to get their names. This operator-based view of functional query languages and methods to implement them are explored in [NIKH84].

The function STotalUnits shown in the database environment is an ordinary function. Here is a possible definition for it:

```
Def STotalUnits s = {   EnrollUnits e = CUnits (^c-enroll e)
                    In
                        fold (+) 0
                            (map EnrollUnits (^s-enroll s))} ;
```

i.e., when applied to a Student, it computes that student's total units using other database functions. This kind of function is sometimes called a "derived function" in the literature.

Here is a recursive query that checks if the course "6.004" is directly or indirectly a prerequisite for the course "6.847":

```
{ q c1 c2 = if (c1 == c2) then true
            else fold (or) false (map (q c1) (CPrereqs c2)) ;

  In
    q (^CName "6.004") (^CName "6.847") }
```

Note that one mixes database and ordinary functions freely. Definitions for ordinary functions may use recursion, conditionals, *etc*. In short, the language is a full programming language.

4.1.2. Operations on Database Functions

For queries, one makes no distinction between ordinary functions and database functions,*i.e.*, one forgets that a database function is a subtype of ordinary functions., and the only interesting operation on functions that we use is function-application. In order to deal with updates, however, there are some additional useful operations on database functions. These operations are *incremental definition* operations, and are inspired by *I-structures* of dataflow research [ANP87b].

If f is a database function (a subtype of t1 -> t2), the expression:

```
domain f
```

has type *t1, and returns a list of objects of type t1 on which f is defined. This is analogous to an operation on an array that returns its index bounds.

If f is a database function (a subtype of t1 -> t2), the expression:

```
new f
```

returns a new database function of the same type that is everywhere undefined. Let this new function be called g. If x:t1, y:t2 and g is undefined at x, then the statement:

```
g [x] = y
```

is an *incremental definition* of g, *i.e.*, it refines, or constrains g so that it now maps domain value x into range value y. By executing many such statements, the meaning of g is gradually "filled in". It is a runtime error to specify two different mappings for g at any domain value x. This distinguishes it from conventional assignment statements, and is sometimes also called the "single-assignment" criterion.

For example, suppose we wish to compute a function that is identical to SStatus except that the status of student Smith is to be special. We could write this (in laborious detail) as:

```
     {
(1)       g = new SStatus ;
(2)       smith = ^Sname "Smith" ;
(3)       g [smith] = "special" ;
(4a)      map (fun s. g[s] = SStatus s)
(4b)           (list_without (domain SStatus) smith) ;
     In
(5)       g }
```

(The line numbers are not part of the program.) In line 1, g is bound to a new undefined function with the same domain and range types as SStatus. In line 2, smith is bound to the student object with name "Smith". In line 3, g is refined to map Smith to "special". In lines 4a and 4b, g is refined so that (g s) is the same as (SStatus s) for all other students.[5] In line 5, g is returned as the value of the whole block.

Note that the incremental definition statement is *not* a destructive assignment. The single-assignment criterion ensures that g never has a value different from its previous one— it always has a value consistent with its previous value. Each incremental definition of g *monotonically* increases its information content.

A dramatic implication of these semantics is that it retains the parallelism and deterministic semantics of a conventional functional language. Unlike imperative languages with destructive assignments, there is no required sequential ordering between the various expressions of a program. In our example above, lines 3 and 4a-4b can be executed in parallel. The function g has a unique meaning, independent of the order of execution.

4.1.3. Update Transactions

An update transaction in FDBPL is a specification of a new database environment e' in terms of the existing database environment e. Recall that an environment is a set of bindings of names to types and values. We follow the convention that a name of the form f' refers to the value of f in e', while a name of the form f refers to the value of f in e.

Semantically, an update transaction in our example database that merely changed student Smith's status to "special" would specify trivial new bindings for all names except SStatus:

---

[5] We use the notation (fun x. e) for lambda-expressions.

```
Student' = Student ;
mkStudent' = mkStudent ;
Students' = Students ;
SName' = SName ;
SStatus' = {g = new SStatus; ... as in example above ... } ;
STotalUnits' = STotalUnits' ;
...
```

Syntactically, of course, this would be too verbose, because in most database updates, the bulk of the database is untouched, and only a few objects are likely to have interesting redefinitions. So, while retaining the above semantics, we provide the following more convenient syntactic conventions for an update transaction:

- Unless explicitly dropped using "drop f", every name f is implicitly carried over to the new environment as if it had been defined by "f' = new f".

- Only new name bindings and interesting incremental definitions are specified in the update transaction.

- At the end of an update transaction, if d' is the domain on which f' is defined (*via* incremental definitions in the transaction), the following expression is executed for each f:

```
map (fun x. f' [x] = f x)
    (list_difference (domain f) d')
```

i.e., for all domain values of f for which f' was not defined, f' is now defined to have the same mapping as f.

Thus, syntactically our example update transaction to change Smith's status to "special" need only say:

```
SStatus' [^Sname "Smith"] = "special"
```

The names from both the new and the old environments are available in the update expression. For example, to increase the weight of the course "Intro to Spelunking" by 3 units, we would write:

```
CWeight' [^CName "Intro to Spelunking"] = (CWeight c) + 3
```

To increase the weight of *all* courses by 3 units, we would write:

```
map (fun c. CWeight' [c] = (CWeight c) + 3)
    Courses
```

4.1.4. A Functional Database System

The above discussion was concerned only with the question of how *one* update transaction specifies a new environment with respect to a given old environment. A Functional Database *system* still has to decide how to connect together multiple updates and queries. There are many possibilities, differing in the kind of serialization of transactions that is necessary.

The simplest possibility is for the system to maintain only one, current database environment. Queries are evaluated in this environment. After computing a new environment in response to an update transaction, the transaction is finally committed by designating this as the current environment and discarding the old one. For consistency, update transactions must be serialized, and multiple queries can be run concurrently, but only between transaction-commits. Even though such a system appears similar to a conventional database system, we will argue in the next section that the use of a functional language itself has advantages.

A second possibility is for the system to maintain a *history* of database environments. Each query or update transaction is evaluated with respect to the environment that is latest at the moment the transaction is received. Each update transaction, when it commits, extends the history with the new environment that it specifies. To maintain a linear history, update transactions must be serialized. Multiple queries may be evaluated concurrently; they never have to be delayed because of update transactions.

In this scenario, one could also extend the language so that expressions can be evaluated in any environment in the history, instead of just the current one. First, we need a sub-language of "environment expressions" to identify specific environments, for example, by an absolute or relative index in the history, or by a real-time stamp. Using this, we can qualify an expression by the environment in which it should be evaluated. The notation used above— f' and f refer to new and old versions of f, respectively— can be seen as a special case of environment-relative naming.

In Section 4.3 we argue that this linear history model is an attractive and useful one.

A third possibility is to maintain a *tree* of database environments rather than a linear history. Here, an update transaction can grow a fresh branch by specifying any point in the tree as the environment relative to which it is to be evaluated. The only transactions that need to be serialized are those that grow the same branch of the tree. Of course, in this tree model, it is necessary to have a way to refer to different points in the tree.

While this third possibility is very general, it is not clear how useful it is— a database being a shared resource, the linear history seems to makes more sense, the latest point in the history being the common, shared view of the database. However, the tree model can be useful, especially in engineering databases, where it is often desirable to try several experimental alternatives to a design, later choosing one of them as the common, shared view.

4.2. Why Functional Languages?

4.2.1. Expressiveness

It is widely accepted that functional languages have great elegance and expressive power [TURN81]. Various high-level features of functional languages give them this expressive power.

In functional languages, functions are first-class values. This is a powerful abstraction mechanism that permits the programmer to design appropriate control structures for existing and new types of objects, thus leading to compact, transparent notation. For example, in defining a new table type in a database, one can simultaneously define general-purpose generators, iterators and reducers to operate on such tables, so that subsequent uses of such tables are concise and clear.

Non-strictness of functions is also a useful tool, making it feasible to define and manipulate infinite and/or large objects perspicuously [TURN81]. However, non-strictness requires that there be no destructive update and its attendant sequentiality, because the order in which expressions are evaluated is in general unpredictable.

Functional languages are expression-based— the main composition rule is function application and is used uniformly from small expressions to large programs. Thus the same language can be used not only for quick, one-off queries in interactive environments but also for large compiled programs.

Even though we believe that functional languages are easier to use because of their clean semantics and high-level features, they are still formal languages. Under certain circumstances, such as casual use of a database by untrained users, one may wish to use other front-ends such as natural-language or graphical interfaces. In such situations, the regularity of functional languages makes them good target languages into which queries are first compiled.

4.2.2. Query Optimizations via Transformation

The semantic cleanliness of functional languages makes it feasible to perform many meaning-preserving program transformations automatically. Because functional languages are referentially transparent, they admit a rich algebra of programs, a feature essential for effective query optimization. The success of current relational database implementations relies in no small measure on the ability to optimize queries in the relational algebra, which is a (restricted) functional language.

4.2.3. Optimizations Due to Parallelism and Non-strictness

Side-effects in a language introduce many read-before-write and write-before-read constraints. Most of these constraints are artificial and make it very difficult, if not impossible, to move away from a purely sequential scheduling of evaluation activities. In functional languages, only the logically necessary data-dependencies remain; this reveals much fine-grained parallelism that permits great latitude in scheduling evaluation activities. This degree of freedom is exploited heavily in dataflow architectures to overcome memory latencies [AI87] [AN87b]. We believe that the same techniques could be used to overcome disk latencies too, a major factor in database system (in)efficiency.

Non-strict functions also increase the parallelism available in a program, because they relax data dependencies even further [AN87b]. In addition, non-strict functions have certain automatic database optimization capabilities in that unnecessary computations (which may involve disk accesses) never get done [NIKH84].

We are currently investigating the possibility of high performance *via* parallelism by studying support for persistence in the MIT Tagged-Token Dataflow architecture.

4.2.4. Cleaner Type Systems

There is much research nowadays into polymorphic type systems [LNCS84]. Polymorphism allows an economic style of programming *à la* Lisp but with complete type safety and with the rich data abstraction facilities necessary for database work. These data abstraction facilities include user-defined abstract types, and type-heirarchies with inheritance.

Much of this research is centered on functional programming languages. Extensions to languages with updates is not often straightforward. For example, in ML there is a well-known problem with polymorphic mutable cells. Consider the following program:

```
def f:(t -> (list t)) =
  {  a = ref nil
     g x = {prog
              a := cons x (deref a) ;
              deref a }
    In
      g } ;
```

Here, f is function with a local variable a, bound to a polymorphic mutable cell, initially containing the polymorphic list nil. In each call, the list is updated by consing the argument onto this list. The function has been declared to have the polymorphic type (t -> (list t)) and its definition is consistent with that declaration. The calls (f true) and (f 0) are therefore individually well-typed, but together they produce a list with mixed types, violating the type discipline. We must point out that there are solutions to this problem, but they are either unintuitive, or they involve *ad hoc* language restrictions (ML originally did not allow polymorphic mutable cells for this reason).

The presence of mutable objects also complicates abstract data type facilities. When building an object of abstract type, one is very careful to ensure that its internal representation cannot be accessed by procedures outside the abstract type definition. Such sharing can make it impossible to guarantee the invariants supposedly maintained by the data abstraction, because it is then possible to update the representation of an abstract object without using one of the allowed procedures, and thus circumvent any integrity checks. This sharing occurs because an abstract object constructor often receives parameters from the outside which it may embed in the new abstract object that it creates. In functional languages, this sharing is safe and so the embedding may be achieved very cheaply (by reference); in languages with side-effects, it may imply expensive and/or excessive *copying* of objects in order to ensure that representations of abstract objects are truly private. In addition to this execution overhead, there is the intellectual overhead for the programmer in that he has to be aware of this pitfall and must explicitly take precautionary measures.

4.3. Why Functional Databases?

Assuming the linear-history model, functional databases "never forget". A database is never modified; instead, it is "updated" by creating a new version that is appended to the history of versions. The versions form a (conceptually) infinite sequence of database environments. Older versions can be named and accessed just like the latest version.

There are many situations where this approach has already been taken, though perhaps not in such a formal sense. For example, many modern file systems (*e.g.*, Vax/VMS) maintain file versions, and this is widely regarded as far superior to no-backup or one-level-backup file systems. Another example is the "Undo" capability now appearing in some programming environments and text editors. Challis also poses arguments in favour of multi-version databases in [CHAL82], where it is claimed that such databases simplify recovery and concurrency algorithms.

Many existing database systems *do* in fact retain all the information contained in previous versions of a database, but only for crash recovery and audit purposes. This is not part of the data model, and the data is not directly accessible to an applications program.

A functional database can offer a superior environment for concurrency control. Queries (read-only transactions) are never delayed, because there is always a latest, committed version that is never going to be changed subsequently. Update transactions do not interfere with queries, because they always build new versions and never interfere with any version currently being examined by queries. Multiple updates, however, still must be serialized (though if two updates work on different parts of the database, they may actually be able to proceed in parallel).

Assuming parallel execution, non-strictness can also improve update performance, because it allows logically serialized operations to be physically overlapped, thus "pipelining" a sequence of updates, instead of doing them strictly one at a time.

Functional databases permit uniform access to historical data. In current databases, if access to historical data is required, the database designer must explicitly encode histories as lists in the schema, and write applications to maintain those lists. An example is the history of withdrawals and deposits against a bank account. Thus, the way these histories are encoded and accessed can vary from database to database, and even within a single database. Further, historical data is available only if anticipated by the database designer. In a functional database system, since the history of versions is visible *via* expressions that are qualified by environment expressions, the programmer has uniform access to the history of any part of the database.

Because the model explicitly supports the notion of a new version of the database, and because it is so easy to revert to old versions, functional databases permit one to write "what-if" programs that experiment with possible futures. This capability would be extremely useful in databases for engineering design, and in so-called intelligent databases or knowledge-bases. For example, when designing a part, there may be several alternative designs that must be pursued simultaneously because it is not yet clear which one would be the most suitable.

Functional databases may ameliorate the difficult problem of manually or automatically merging two separate databases into one. This problem arises, for example, when the operations of two or more companies are consolidated. Typically, the data in the two source databases will disagree in various ways. Having access to the histories of the two databases may alleviate the inconsistency problem because one may be able to use pattern-matching algorithms to establish consistent correspondences between the data at different points in the history of the two databases.

In many information systems, permanent access to the history of the database is important. It may be a legal requirement to maintain records of all transactions, for example in a bank or personnel database (in fact banks already do this, but without any direct help from the data model or the database management system). The ability to examine time trends, for example in a stock-market or econometric modelling database, can be an invaluable capability.

5. IMPLEMENTATION ISSUES

We are in the process of designing a prototype of a Functional Database System. Here are some of our early design ideas.

There are many implementation techniques currently in use for functional languages. The one that we are currently pursuing is the dataflow approach [ANP87b], because of its support for non-strictness and because it is an approach to parallelism for higher performance.

To implement an FDBPL, the memory is organized into a sequence of heaps. Each version of the database is associated with its own heap, though this does not mean that the entire database version resides on that heap. Each update transaction initiates a new heap, and new objects constructed as part of that transaction are allocated in the new heap, though they may refer back to objects in older heaps. Thus an updated version of the database shares much of the previous versions; only new objects allocated as part of the new version initially reside in the new heap for that version.

The programmer's model of memory is that the entire sequence of databases (*i.e.*, the sequence of heaps) is implemented in stable storage and is resilient across crashes. As a transparent optimization, pages of various heaps will have temporary working copies in high-speed, volatile memory. When a transaction is committed, the copy in stable storage is made to agree with any extant temporary copy.

Even though the data model presents the database as a potentially infinite history of versions, because of finite storage capacity it will in general be necessary to prune the sequence of databases at some version, and move all prior versions off-line. To do this, all data shared with prior versions are copied forward into a fresh heap at the oldest retained version. Decisions to prune the version history, and identification of the oldest version to be retained are taken dynamically depending on current storage utilization, for example by a database administrator.

Our semantics ensures that one cannot see different data as a result of this pruning activity. When the version history is pruned, one will no longer be able to execute a query that interrogates a pruned version— such attempts will invoke a run-time error. Despite this, if the total storage capacity of the database system is subsequently increased (*e.g.*, by adding more disks), it is possible to re-integrate earlier pruned verions easily and smoothly.

We emphasize though we still have to live with finite databases, we believe our model is still a step forward. We have relieved the database programmer of concerns about finiteness in much the same way that the Algol programmer is relieved of certain finite storage allocation issues that bedevil the Fortran programmer, even though in a deep theoretical sense they are certainly equivalent.

It appears that our model seems naturally suited to high-capacity write-once storage technology such as optical disks, but we have yet to study algorithms that make effective use of such devices in this functional setting.

6. CONCLUSION

We have argued that it is essential that databases be treated as an integral part of programming languages rather than separately, as is common today. Further, we have claimed that the full range of type-structures in a programming language should be available for database objects. We are in substantial agreement with other researchers on these positions.

We then put forward the view that functional languages offer many advantages having to do with expressive power, rich type systems, interactive environments, parallelism and optimization. Databases should be viewed as environments in which queries (*i.e.*, expressions) are evaluated. They should be "updated" not by destructive assignment, but by specifying a new environment, thus creating a conceptually infinite sequence of database versions. We call such a system a Functional Database Programming Language. We argue that FDPLs are superior vehicles for concurrency control, and that existence of a model of the history of a database is useful not only for recovery mechanisms but also in general for application programs.

We have begun developing a prototype functional database programming language to demonstrate the feasibility of this model.

REFERENCES

[ABCCM81] [ABCCM84] [ACO85] [AGOP85] [AI87] [ANP87a] [ANP87b] [BMS84] [CARD83b] [CHAL82] [DATE86] [GMN84] [LNCS84] [NIKH84] [NIKH87] [SCHM77] [TURN81]

Chapter 6

A New Type-Checker for a Functional Language

Jon Fairbairn
University of Cambridge

ABSTRACT   *The polymorphic type system of the programming language Ponder [FAIR82] is described. The initial sections give an overview of the syntax of Ponder, some of the motivation behind the design of the type system, and a description of its relationship with other type systems. This is followed by a definition of the relation of 'generality' between Ponder types, and of the notion of type-validity of Ponder programs. An algorithm to determine whether a Ponder program is type-valid is then presented. The final sections give examples of useful types that may be constructed within the type system, and describe some of the areas in which it is thought to be inadequate.*

1. INTRODUCTION

This[1] is a description of the type system of the pure functional programming language Ponder [FAIR82]. Ponder was designed to demonstrate that a simple, conceptually clean language can be made practical, and for experiments in functional programming style. To these ends the language contains no built-in data-types or type constructors, no built-in flow control primitives and no assignable variables.

Despite this austerity of design Ponder has been implemented to a level of efficiency suitable for moderately large programs. The largest program to date is a novel spreadsheet system with a built in concept of the passage of time (implemented by my colleague S. Wray). The spreadsheet is sufficiently powerful to allow the simulation of small sequential logic circuits, and amounts to some 2500 lines of Ponder. The experience of writing this program has demonstrated the efficacy of the type-checker: most of the time writing it was spent correcting type-errors, and when it was passed by the type-checker, it was a working system.

The purpose of type-checking is to detect errors before the program runs, but what is classed as a type-error is a matter of choice. For example, it is possible to express as type information the fact that the reciprocal function only works for non-zero arguments. In practice such a restriction would result in programs cluttered with redundant tests for zero, and the reciprocal function is given a type that includes zero. Conversely the programr may want to use a subset of the integers to represent some other object. In this case it is useful to define them as a different type to ensure the integrity of the program. Type-checking can assure us that no function is applied to an argument outside its *intended* domain. The Ponder type-system allows this freedom of choice. Indeed, it is possible to give expressions such types that the application of any one to any other is

---

[1]   An earlier version of this paper appeared in Science of Computer Programming 6 (1986) pp 273-290. The present version corrects some errors in the description of the type-checking algorithm.

type-valid (fortunately expressions typed in this way cannot be applied to expressions with more sensible types).

The strong typing of languages like Pascal or Algol68 is too restrictive. The append function for lists works independently of the type of the elements of the lists, yet in either of these languages it must be rewritten for every type of list. The solution to this problem is polymorphism, as in ML [GMW79] [MILN78] in which types may include type variables. The system in ML is too weak to allow the definition of some useful type-constructors (Cartesian product, for example), and although these could be added as primitives this would be contrary to the spirit of Ponder. Ponder types also have the advantage over ML that they allow the use of polymorphic arguments in functions. The polymorphism in Ponder is closer to that of MacQueen and Sethi [MS82] [MS84] but includes fewer built-in constructors and is restricted to make mechanical type-checking possible.

Although I give formal rules for type-checking, my intention in this paper is to present the type-system as a practical tool rather than for its theoretical significance, so I feel that it is excusable not to have included proofs of soundness or completeness.

2. RELATIONSHIPS WITH OTHER TYPE DISCIPLINES

It will be instructive to examine the relationship between the Ponder type system and others. There are two main kinds of polymorphism: object- and type- polymorphism[2]. In object-polymorphism a type is regarded as being determined by the set of operations that can be performed on objects of the type. Russel [DD80a] and in a sense Poly [MATT85b] are examples of languages with type systems of this sort. Ponder has type-polymorphism, so it is more closely related to languages such as ML.

In type-polymorphism the types themselves express the polymorphism, by including *type-variables*. Consider the function

```
λa.λf.f a
```

Without polymorphism we must choose a type such as

$$\text{Int} \rightarrow (\text{Int} \rightarrow \text{Int}) \rightarrow \text{Int}$$

or

$$\text{Char} \rightarrow (\text{Char} \rightarrow \text{Char}) \rightarrow \text{Char}.$$

Such types might be written something like

```
proc(Int)proc(proc(Int)Int)Int
```

in more traditional type systems, but the notation used here is more economic, and corresponds more closely with mathematical tradition. Obviously the function works just as well whether the type is `Int` or `Char`, and the idea is to replace the irrelevant details with a type-variable that can be filled in later. So the function might now be given the type $\alpha \rightarrow (\alpha \rightarrow \alpha) \rightarrow \alpha$.

---

[2]   I have used these terms for want of any better.

The presence of this variable might lead one to ask where it is declared, and how does it get instantiated? In ML these questions are answered behind the scenes --- in effect type-variables are declared wherever they are used, and instantiated by the compiler whenever necessary. This approach has the advantage that the programr may omit almost all type information, the compiler inferring it from context. The disadvantages are that the question of scope of a type variable becomes a little delicate, certain kinds of polymorphism are inexpressible and the Principle of Correspondence is not adhered to. The principle means that a program in which an expression is given a name should be the same as a program in which the expression is passed as a parameter to the rest of the program. In ML declarations can be polymorphic but parameters cannot be. Another example will illustrate this: the expression

$$(\lambda f.f\ 3,\ f\ \text{``hello''})\ (\lambda x.x)$$

has no type in ML because the bound variable f is applied to two different types of argument, whereas the corresponding expression

$$\textbf{let } f \triangleq \lambda x.x \textbf{ in } f\ 3,\ f\ \text{``hello''}$$

has.

There are two ways of solving the problem of binding type variables and both of them at least partially solve the problem of correspondence. One way is to bind the type variables by abstracting over them. This is the approach adopted in the Second order typed lambda-calculus [REYN86]. Here type variables are bound with a Δ, and the programr must supply values for them whenever a polymorphic function is used. The type of example 1 is now written Δ t.t $\rightarrow$ (t $\rightarrow$ t) $\rightarrow$ t, but whenever it is used it must be given a type for t. Here the problem of correspondence is solved by making the argument f in example 2 polymorphic, so that the type of the function can be written (Δ T.T $\rightarrow$ T) $\rightarrow$ Int X String. Again f must be supplied with a type whenever it is used.

The other approach to the binding of type variables, the one adopted for Ponder and in the MacQueen/Sethi system, is to quantify over them. The type of the function in example 2 might now look like this: ($\forall$T.T $\rightarrow$ T) $\rightarrow$ Int X Char, but this time there is no need to supply type arguments to f.

In Ponder the principle of correspondence is not adhered to, but in a minor way. It is possible to write an expression that has more than one type, such that the type checker cannot decide which type is intended until the expression is used. If such an expression is bound with a Let declaration, it can be used in more than one situation, so that more than one of the types is needed, and of course there is no way of doing this to a bound variable. There are two ways in which Ponder could be altered to conform to the principle. The type checker can check to see that all occurrences of such expressions are used with the same type, or can insist that the programr states which type is meant in the declaration. Neither alternative seems worthwhile, since the former is impossible for separately compiled sections, and the latter just gives the programr more work.

The three type-polymorphic disciplines above differ considerably in respect of the complexity of type checking. In the ML type system every expression that has a type has a most general type, which is to say that there is a type that subsumes all other types that the expression can have. This, together with the relative simplicity of the types makes it possible to infer the types of expressions algorithmically using unification.

Type checking for the second order typed lambda-calculus is simpler, in that all the necessary type information is supplied by the programr, so that the checker need only compare types for equivalence. For the Ponder type system it is much more difficult, because the values of type variables are not supplied by the programr, and expressions do not have most general types. The algorithm presented below does not infer the values for type variables. It instead places limits on the range of values that the variables can take, and hence on the ranges of types.

3. SYNTAX

This section describes the syntax of types and of the kinds of Ponder expression that are relevant to the discussion of type-checking.

3.1 Types

Ponder types are constructed from function types, quantified types and type generators:

$$\text{Type} \;=\; \begin{cases} \text{V} & \text{Type variables} \\ \text{Type} \to \text{Type} & \text{Functions} \\ \forall \text{V}.\text{Type} & \text{Quantified Types} \\ \text{G[Type,\ldots,Type]} & \text{Generators} \end{cases}$$

"Generators" are user defined parameterised type constructors. A further facility is the capsule, which provides a very simple kind of encapsulation of types.

Note that there are no built-in types at all. This is unimportant since parameterless capsules that behave like built-in types (such as `Integer`) can be constructed. Capsules with parameters perform a similar function to that of built-in type constructors.

3.1.1 Naming

For clarity I will give types names that are either a single upper-case letter (e.g. `T`), or are words beginning with an upper-case letter followed by letters, hyphens and possibly with a subscript (like `T`, T_1, `Long-name`, ...).

3.1.2 Function Types

The most primitive kind of type is that of the function from one type to another, which is written using "$\to$". Thus if `Parameter` and `Result` are both types, then `Parameter` $\to$ `Result` is the type of a function taking objects of type `Parameter` to objects of type `Result`. Note that $\to$ associates to the right, so that `A` $\to$ `B` $\to$ `C` means the same as `A` $\to$ `(B` $\to$ `C)`.

3.1.3 Quantifiers

The universal quantifier, ∀, (pronounced 'for all'), introduces a name within the rest of a type or expression. Hence ∀V.V → V is a type, which means 'For all types V, take an object of type V, to another object of type V.' This is the type of the identity function λx.x.

A note about binding: the scope of a variable introduced by a quantifier extends as far to the right as possible, but is limited by parentheses, so ∀T.(T → T) → Bool means the same as ∀T.((T → T) → Bool), and takes as argument any function with the same parameter and result types, whereas (∀T.T → T) → Bool demands that its argument has type ∀T.T → T (and hence could not be expressed in Milner's type system). For the sake of convenience, ∀T.∀U... may be written ∀T,U...

3.1.4 Type Generators

The final kind of type constructor is the type generator, which is a user defined operator that generates a type, for example:

> Type Identity ≜ ∀V.V → V;

This declares (≜ means 'is defined as') a generator Identity, which means the same as ∀V.V → V.

Generators may also have parameters:

> Type Arrow[Left,Right] ≜ Left → Right

where Left and Right are the names of the parameters, so that
> Arrow[Bool,Bool]

means the same as
> Bool → Bool,

and
> Arrow[Int,Real]

is the same as
> Int → Real,

and so on. For example, we might define pair types as

> Type Pair[L,R] ≜ ∀Res.(L → R → Res) → Res

Note that this is an example of a type that is not expressible in ML. (This type is described in more detail in the section 'Representing Objects' below.)

Finally Ponder allows types to be recursive (but in a restricted way); so we can have declarations like

> Rectype Infinite-List[T] ≜ Pair[T,Infinite-List [T]]

which means that Infinite-List[Int] means the same as Pair[Int,Pair[Int,Pair[Int,...]]].

Recursive generators are restricted to ensure that all types generated are finite cycles rather than infinite trees. In a definition

$$\texttt{Rectype G[T}_1, \dots, \texttt{T}_n] \triangleq \texttt{Body}$$

all applications of G within Body must be to *exactly* the parameters [$\texttt{T}_1, \dots, \texttt{T}_n$]. One way of looking at this restriction is to consider it in terms of a μ-*operator* for type generators. μ binds a type variable (*not* a generator) and finds a fixed point of a type generator, so that if we were to allow type definitions like

$$\texttt{Type G[X]} \triangleq \mu\texttt{T.T} \rightarrow \texttt{X}$$

then G would be the same as if it were defined as in

$$\texttt{Rectype G X]} \triangleq \texttt{G[X]} \rightarrow \texttt{X.}$$

Recursive types in Ponder are just those types constructible with the μ-operator. For example:

$$\texttt{Rectype Invalid[T]} \triangleq \texttt{Invalid[(T} \rightarrow \texttt{T)]}$$

is invalid. This restriction is necessary to make mechanical compile-time type checking possible; without it type generators would be equivalent to the λ-calculus, and hence type-checking would be undecidable, since comparison of types would be equivalent to computation of the equality of functions.

3.1.5 Capsules

Normally, two differently named definitions of the same type are equivalent; i.e. are indistinguishable in use (the exact rules for equivalence are given below). For example, if

$$\texttt{Type A} \triangleq \forall\texttt{T}_1.\texttt{T}_1$$

and

$$\texttt{Type B} \triangleq \forall\texttt{T}_1.\forall\texttt{T}_2.\texttt{T}_2$$

A and B may be used interchangeably since they define equivalent types. This is not always what is wanted, so capsules are provided as a means of destroying the referential transparency of a particular name. Capsules are the same as generators, except that the structure of the declared type is hidden for the purposes of type-checking. This means that after the declaration

$$\texttt{Capsule-type P} \triangleq \texttt{Int}$$

P is considered to be a different type from Int. Thus capsules obey 'name equivalence' (as do abstract types in ML), whereas types in general obey the rules of equivalence given below, in a similar way to the structural equivalence of types in Algol 68 [VANW75]

3.2 Expressions

$$
\text{Expression} = \begin{cases}
\texttt{Name} & \text{Variable} \\
\texttt{Expression}_1 \ \texttt{Expression}_2 & \text{Application} \\
\texttt{Type Name} \rightarrow \texttt{Expression} & \text{Abstraction} \\
\texttt{Type:Expression} & \text{Cast} \\
\forall \texttt{V.Expression} & \text{Quantified Expression}
\end{cases}
$$

3.2.1 Grouping

Parentheses are used merely to achieve a particular binding, but square brackets are used around lists of arguments to type generators.

3.2.2 Function Application

The syntax of function application is the same in Ponder as in the λ–calculus: `f x` means 'apply `f` to `x`'. Application associates left, so `f x y` means `(f x) y`

3.2.3 Abstraction

Ponder uses a compact method of specifying the name and type of the bound variable of a function:

$$\texttt{Type Name} \rightarrow \texttt{Body}$$

is a function with bound variable `Name` that is required to have type `Type`. The equivalent λ-function would be `λName.Body`.

For example

$$\texttt{Int i} \rightarrow \texttt{plus i 1}$$

is a function taking an argument of type `Int` called `i`, and returning the value of `plus` applied to `i` applied to `1` (i.e. λ `i. plus i 1` in the λ–calculus).

3.2.4 Casts

A cast is an expression that is asserted by the programr to have a particular type:

$$\texttt{Type : Expression}$$

means that `Expression` must have type `Type`.

3.2.5 Quantified Expressions

An expression may be preceded by a type quantifier:

$$\forall V . \texttt{Expression}$$

This declares the type variable v in the expression and imposes the constraint that the expression be valid whatever type is substituted for v.

So in Ponder the identity function can be written

$$\forall V . V \ x \ \rightarrow \ V \ : \ x$$

having the type `Identity`.

4. TYPE VALIDITY

This section introduces the rules to which a Ponder program must adhere.

Although it is not mentioned above, Ponder allows the definition of *overloaded* operators (operators for which the meaning is dependent upon the types of the arguments). This means that it is impossible to infer the type of an expression unless the types of all the variables are known. However, Ponder is similar to HOPE [BMS80] in the respect that the syntax ensures that all variables are declared with a particular type.

On the other hand, it would be tedious to have to give the result types of every function application, because in most cases it is possible to infer the result type from the types of the function and argument. Hence the Ponder type-checker must be capable of determining whether an expression is type correct, given the types of all the variables present.

In order to describe this, I need to give a definition of the relationship of generality between types. The relation $T_1 \geq T_2$ is intended to mean that any object of type T_1 may validly be used in any situation where an object of type T_2 may validly be used. Thus, if `Identity` is as defined above, `Identity` $\geq$ (`Int` $\rightarrow$ `Int`), since any object of type `Identity` (they are all equivalent to $\lambda x . x$) may safely be used where an object of type `Int` $\rightarrow$ `Int` may be used.

As programs contain sub-expressions with type variables defined at an outer level, it is necessary to consider the possibility of comparison of types containing free variables. For example, within the body of the function $\forall V . V \ x \ \rightarrow \ ...$, the parameter x will be of type v. Free type variables like this can be treated as constant types, because there is no information available about the type that they represent.

Capsules with no arguments are treated in exactly the same way as type variables bound at an outer level. Capsules with arguments require slightly more complex treatment, since we want to ensure that applications of the same capsule to different arguments compare in a friendly way. If `Arrow` were a capsule, we would still want `Arrow [Int, Identity]` $\geq$ `Arrow [Int, (Int` $\rightarrow$ `Int)]`.

To achieve this, when comparing two capsules, first discover whether they are the same capsule (if not, then they are incomparable), and then compare the types resulting from the application of the capsule to its two sets of arguments.

4.1 The Relation of generality between types

Rules R0 to R8 below define the relation $\geq$. V_n are type variables, T_n are arbitrary types (possibly with free variables), G_n are generators, Γ stands for a set of assumptions each of which is of the form $T_1 \geq T_2$ or $G [V_1, ..., V_2] \triangleq T$ and $\geq$ is as above.

Assumption

$$\Gamma \cup \{ T_1 \geq T_2 \vdash T_1 \geq T_2\} \qquad\qquad \text{R0}$$

Reflexivity

$$\Gamma \vdash T \geq T \qquad\qquad \text{R1}$$

This means that from any set of assumptions Γ and type T we can deduce that $T \geq T$ i.e. any type is at least as general as itself.

Transitivity

$$\frac{\Gamma_1 \vdash T_1 \geq T_2, \quad \Gamma_2 \vdash T_2 \geq T_3}{\Gamma_1 \cup \Gamma_2 \vdash T_1 \geq T_3} \qquad\qquad \text{R2}$$

In rules such as this, assumptions are written above the line and conclusions below it. This rule may be read as 'If we can prove from Γ_1 that $T_1 \geq T_2$ and from Γ_2 that $T_2 \geq T_3$, then we can prove from the union of Γ_1 and Γ_2 that $T_1 \geq T_3$.'

Instantiation

$$\Gamma \vdash \forall V.T_1 \geq T_1[T_2/V] \qquad\qquad \text{R3}$$

(Expressions of the form $T_1 [T_2/V]$ mean "T_1 with every free occurrence of V replaced by T_2.") An object that works for all types is more general than an instance of it.

Generalisation

$$\frac{\Gamma \vdash T_1 \geq T_2}{\Gamma \vdash T_1 \geq \forall V.T_2} \quad V \text{ not free in } T_1 \text{ or } \Gamma \qquad \text{R4}$$

If a type T_1 is more general than a type parameterised on V, regardless of the value of V, then T_1 is also more general than the generalised version of that type.

Function

$$\frac{\Gamma_1 \vdash T_3 \geq T_1, \ \Gamma_2 \vdash T_2 \geq T_4}{\Gamma_1 \cup \Gamma_2 \vdash T_1 \to T_2 \geq T_3 \to T_4} \qquad \text{R5}$$

A function that requires a less general argument is more general. This is explained by the following analogy. If a person is giving you something you want, the more he gives, the more generous he is. Conversely, if a person is taking something you owe him, the *less* he takes the more generous he is.

Result

$$\Gamma \vdash (\forall V.T_1 \to T_2) \geq T_1 \to \forall V.T_2 \quad V \text{ not free in } T_1 \qquad \text{R6}$$

A quantifier that does not appear in the parameter specifier of a function can be moved to the result. In fact the reverse is also true; see **3.2 Properties** below.

Recursion

$$\frac{\Gamma \cup \{G[V_1, ..V_n] \triangleq T, G[T_1, ..T_n] \geq T_0\} \vdash T[T_1, ..T_n/V_1, ..V_n] \geq T_0}{\Gamma \cup \{G[V_1, ...V_n] \triangleq T\} \vdash G[T_1, ...T_n] \geq T_0} \qquad \text{R7}$$

this rule allows the comparison of recursive types.

Expansion

$$\Gamma \cup \{G[V_1, ...V_n] \triangleq T\} \vdash T[T_1, ...T_n/V_1, ...V_n] \geq G[T_1, ...T_n] \qquad \text{R8}$$

this gives the meaning of definition (the case with the generator on the left hand side of $\geq$ is covered by R7).

4.2 Properties

The type checker makes tacit use of some straightforward consequences of these rules including the following:

- $\forall V_1, V_2.T \ \ddot{Y} \ \forall V_2, V_1.T$
  (where $T_1 \equiv T_2$ means $T_1 \geq T_2 \wedge T_2 \geq T_1$, that is T_1 is equivalent to T_2), i.e. the order in which quantifiers appear in a type is irrelevant to its meaning. (By R3 and R4).

- $\forall V_1, ...V_n, V.T \to G[V] \ \ddot{Y} \ \forall V_1...V_n.T \to \forall V.G[V]$
  Provided that V is not free in T (By R6, R3 and R4.) Hence types may be equivalent even when they have different textual representations.

- $\Gamma \vdash (\forall V.V) \geq T$ for all T (By R3).

An object of type $\forall V.V$ has every type. The only such object is the non-terminating computation.

4.3 Rules for Type-Validity of Expressions

This section presents the rules to which valid Ponder programs must conform. In general a program will consist of a "casted" expression, the type of which is determined by the environment in which the program is intended to run. However, a program without a cast at the top level must be type-valid; that is, the program must have at least one type.

A program p is type-valid if a statement of the form $T:p$ for some T may be proved within the following rules.

The notation $T:e$ means that e has the type T, and Γ is a set of assumptions as before but may also include assumptions of the form $T:v$.

Variable Assumption

$$\Gamma \cup \{T:v\} \vdash T:v \qquad\qquad\qquad\qquad V0$$

Application

$$\frac{\Gamma_1 \vdash (T_1 \rightarrow T_2):e_1, \ \Gamma_2 \vdash T_1:e_2}{\Gamma_1 \cup \Gamma_2 \vdash T_2:e_1 \ e_2} \qquad\qquad V1$$

Function

$$\frac{\Gamma \cup \{T_1:v\} \vdash T_2:e}{\Gamma_1 \vdash T_1 \rightarrow T_2:(T_1:v \rightarrow e)} \quad \text{where } \Gamma = \Gamma_1 - \{T:v \mid T \text{ is a type}\} \qquad V2$$

Cast

$$\frac{\Gamma \vdash T_1:e}{\Gamma \vdash T_1:(T_1:e)} \qquad\qquad\qquad V3$$

Generalisation

$$\frac{\Gamma \vdash T:e}{\Gamma \vdash \forall v.e} \quad v \text{ not free in } \Gamma \qquad\qquad V4$$

Restriction

$$\frac{\Gamma_1 \vdash T_1:e_1, \ \Gamma_2 \vdash T_1 \geq T_2}{\Gamma_1 \cup \Gamma_2 \vdash T_2:e} \qquad\qquad V5$$

5. THE TYPE CHECKING ALGORITHM

Having established the meaning of type-validity, I now describe an algorithm for type-checking Ponder programs. Note that the algorithm does not infer the types of expressions; instead it produces a type that may contain variables, together with conditions on them---this effectively represents a set of possible types for the expression. To see why this is necessary, consider the application of a function of type $\forall \mathbf{V}. (\mathbf{V} \to \mathbf{V}) \to \mathbf{V} \to \mathbf{V}$ to an argument of type $\mathbf{Int} \to \forall \mathbf{T}.\mathbf{T}$. This argument can be regarded as having either the type $\mathbf{Int} \to \mathbf{Int}$ or $(\forall \mathbf{T}.\mathbf{T}) \to \forall \mathbf{T}.\mathbf{T}$. In the first case, the result of the application will have type $\mathbf{Int} \to \mathbf{Int}$, and in the second $(\forall \mathbf{T}.\mathbf{T}) \to \forall \mathbf{T}.\mathbf{T}$, but neither of these types is more general than the other, so the checker must wait until later before deciding which was intended.

The algorithm is presented as a function *type-check*, followed by definitions of subsidiary functions. For the sake of clarity, the mechanism that deals with recursion is described separately. Where the definition of a function is by case analysis on types, I have omitted the type-generator case since, in the absence of recursive types, all type generators may be expanded out in full.

In what follows, things represented by $\mathbf{V}$ are type variables, $\mathbf{T}$ are arbitrary types and e are Ponder expressions. Whenever $\forall \mathbf{V}.\mathbf{T}$ occurs, it is assumed that it is distinct from all other variables encountered (in practice this is implemented by renaming).

Furthermore, I assume that all types have been normalised so that they contain no redundant quantifiers and so that quantifiers that do appear are moved as far to the right as possible (via R6).

5.1 *type-check*

The type checker *type-check* takes a triple $(\mathbf{A}, \tau, e)$ where e is a Ponder expression, $\mathbf{A}$ is a set of assumptions about type variables and τ is a set of typings of the form $\mathbf{T}: e$. Each element of $\mathbf{A}$ is one of

- Fixed $\mathbf{V}$,
- $\mathbf{T} \geq \mathbf{V}$,
- $\mathbf{V} \geq \mathbf{T}$.

The result of *type-check* $(\mathbf{A}, \tau, e)$ is a pair $(\mathbf{A}', \tau)$ where τ represents the type of e under the new assumptions $\mathbf{A}'$ (which will contain $\mathbf{A}$), if e is valid, and Fail otherwise. In other words, *type-check* $(\mathbf{A}, \tau, e) = (\mathbf{A}', \tau)$ implies that $\mathbf{A}' \cup \tau \vdash \mathbf{T}:e$.

The type checker requires two subsidiary functions *valid*, and ⊇; *valid* checks a set of constraints and either returns the same set of constraints or fails, and $\mathbf{T}_1$ ⊇ $\mathbf{T}_2$ is the set of assumptions needed to show that $\mathbf{T}_1 \geq \mathbf{T}_2$.

The definition of *type-check* is given as a case analysis of possible expressions, together with a short description of the idea underlying each particular case.

Variables

The type of a variable is given by its environment:

$$\textit{type-check } (\mathbf{A}, \tau, v) = (\mathbf{A}, \mathbf{T}), \text{ if } (\mathbf{T} : v) \in \tau \text{ otherwise Fail}$$

Application

To check an application, check the function and argument, and then calculate the result type (cf V1):

$$type\text{-}check \ (A, \ \tau, \ e_1 \ e_2) \ = \ (valid \ (A_1 \ \cup A_2 \ \cup \ T_1 \ \boxed{\geq} \ (T_2 \ \to \ V_r)), V_r),$$
$$where \ \ (A_1, T_1) = type\text{-}check \ (A, \tau, e_1)$$
$$and \ \ (A_2, T_2 \ ) \ = \ type\text{-}check \ (A, \ \tau, \ e_2)$$
$$and \ V_r \ \text{is not free in} \ A_1, A_2, T_1, T_2 \ \text{or} \ \tau$$

Function

Type-check the body of the function given that the parameter has the stated type; the result type is a function from the parameter type to the type of the body (cf V2):

$$type\text{-}check \ (A, \tau, T_v : v \ \to \ e) \ = \ (A', T_v \ \to \ T_e),$$
$$where \ (A', T_e) \ = \ type\text{-}check \ (A, \tau \cup \{T_v : v\}, e)$$

Casts

The type of a cast expression is the type given in the cast, but we must check to see that the type of the expression $\geq$ that type (cf V3):

$$type\text{-}check \ (A, \ \tau, \ T : e) \ = \ (A_2, T) \ where \ A_2 \ = valid \ (\ A_1 \cup \ (T_1 \ \boxed{\geq} \ T))$$
$$where \ (A_1, T_1) \ = type\text{-}check \ (A, \ \tau, \ e)$$

Quantified Expressions

The body of a quantified expression is checked given that the type variable is fixed (cf V4):

$$type\text{-}check \ (A, \ \tau, \ \forall V . e) \ = \ (A' \ - \ \{Fixed \ V\}, \ \forall V . T_e),$$
$$where \ (A', T_e) \ = type\text{-}check \ (A \cup \{Fixed \ V\}, \ \tau, \ e)$$

5.2 $\boxed{\geq}$

$\boxed{\geq}$ is an infix operation between two types. $T_1 \ \boxed{\geq} \ T_2$ is a set of assumptions A, such that $A \vdash T_1 \ \geq \ T_2$. The assumptions may not be consistent; see *valid* below.

The following rewriting rules define $\boxed{\geq}$ case by case (a $\Rightarrow$ b, c means reduce a to b, otherwise try c):

$$T \ \boxed{\geq} \ V \ \Rightarrow \ \{T \geq V\}, \qquad\qquad C0$$
$$V \ \boxed{\geq} \ T \ \Rightarrow \ \{V \geq T\}, \qquad\qquad C1$$
$$\forall V . T_1 \ \boxed{\geq} \ T_2 \ \Rightarrow \ T_1 \ \boxed{\geq} \ T_2 \qquad\qquad C2$$

This corresponds to R3

$$T_1 \ \to \ T_2 \ \boxed{\geq} \ \forall V . T_3 \ \Rightarrow \ (T_1 \ \to \ T_2 \ \boxed{\geq} \ T_3) \ \cup \ \{Fixed \ V\}$$

the "Fixed" represents the fact that V must be free in T_3 as in R4

$$T_1 \ \to \ T_2 \ \boxed{\geq} \ T_3 \ \to \ T_4 \ \Rightarrow \ (T_3 \ \boxed{\geq} \ T_1) \ \cup \ (T_2 \ \boxed{\geq} \ T_4) \qquad C5$$

This corresponds to R5

5.3 Checking Assumption Sets

I now give the rules for *valid*; *valid* first forms the closure of the assumption set and then checks it for consistency. A is a set of assumptions as for *type-check*, and ☑ is defined below.

$$valid = check(\ closure\ A\ ) \hspace{4cm} \text{S1}$$

where *closure* A is the least set C such that

$$A \subseteq C \hspace{6cm} \text{S2}$$
$$\{T_1 \geq V,\ V \geq T_2\} \subseteq C \Rightarrow T_1\ \boxed{\geq}\ T_2 \subseteq C \hspace{1.5cm} \text{S3}$$
$$\{T_1 \geq V,\ T_2 \geq V\} \subseteq C \Rightarrow T_1\ \boxed{\vee}\ T_2 \subseteq C \hspace{1.5cm} \text{S4}$$

Here ⇒ stands for logical implication. For non-recursive types this set is clearly finite, because all types introduced to C are subtypes of types in A. This permits an algorithmic implementation, but care must be taken not to compute the same comparisons repeatedly. The easiest solution to this is to use the same memory mechanism as for recursive types (see **5.6** below).

$$check\ A = \ \text{Fail,} \hspace{1cm} \text{if } \{\text{Fixed } V,\ T \geq V\} \subseteq A \hspace{1cm} \text{S5}$$
$$\text{Fail,} \hspace{1cm} \text{if } \{\text{Fixed } V, V \geq T\} \subseteq A \hspace{1cm} \text{S6}$$
$$A \hspace{1.5cm} \text{Otherwise} \hspace{3cm} \text{S7}$$

In S5 and S6 T is understood to be distinct from V.

5.4. ☑

This operation is reminiscent of unification [ROBI65]. T_1 ☑ T_2 is the set of restrictions on variables in T_1 and T_2 needed to show that there is a T_3 such that $T_1 \geq T_3$ and $T_2 \geq T_3$.

$$V_1\ \boxed{\vee}\ V_2 \hspace{0.5cm} \Rightarrow \hspace{0.5cm} \{V_1 \geq V_2,\ V_2 \geq V_1\}, \hspace{1cm} \text{U0}$$
$$V\ \boxed{\vee}\ T \hspace{0.5cm} \Rightarrow \hspace{0.5cm} T\ \boxed{\geq}\ V, \hspace{2cm} \text{U1}$$
$$T\ \boxed{\vee}\ V \hspace{0.5cm} \Rightarrow \hspace{0.5cm} T\ \boxed{\geq}\ V, \hspace{2cm} \text{U2}$$
$$\forall V.T_1\ \boxed{\vee}\ T_2 \hspace{0.5cm} \Rightarrow \hspace{0.5cm} T_1\ \boxed{\vee}\ T_2, \hspace{1.5cm} \text{U3}$$
$$T_1 \to T_2\ \boxed{\vee}\ \forall V.T_3 \hspace{0.5cm} \Rightarrow \hspace{0.5cm} T_1 \to T_2\ \boxed{\vee}\ T_3, \hspace{0.5cm} \text{U4}$$
$$T_1 \to T_2\ \boxed{\vee}\ T_3 \to T_4 \hspace{0.5cm} \Rightarrow \hspace{0.5cm} T_2\ \boxed{\vee}\ T_4 \hspace{1.5cm} \text{U5}$$

Notice that in U5 no account is taken of T_1 or T_3 because the type $\forall V.V \to T$ is always less general than both $T_1 \to T$ and $T_3 \to T$.

5.5. Examples

If f and x have been declared with types $(\text{Int} \to \text{Int}) \to \text{Bool}$ and $(\forall T.T \to T)$ respectively, then *type-check* $(\{\}, \{(\text{Int} \to \text{Int}) \to \text{Bool}:f, (\forall T.T \to T):x\}, f\ x)$ results in checking f and x, which in turn results in

$$(valid \; ((\texttt{Int} \; \rightarrow \; \texttt{Int}) \; \rightarrow \; \texttt{Bool} \; \boxed{2} \; (\forall \texttt{T.T} \rightarrow \; \texttt{T}) \; \rightarrow \; \texttt{V}_r) \, , \texttt{V}_r)$$

taking

$$(\texttt{Int} \; \rightarrow \; \texttt{Int}) \; \rightarrow \; \texttt{Bool} \; \boxed{2} \; (\forall \texttt{T.T} \; \rightarrow \; \texttt{T}) \; \rightarrow \; \texttt{V}_r$$

we get

$((\forall \texttt{T.T} \; \rightarrow \; \texttt{T}) \; \boxed{2} \; \texttt{Int} \; \rightarrow \; \texttt{Int} \; \cup \; (\texttt{Bool} \; \boxed{2} \; \texttt{V}_r)$	By C5
$(\forall \texttt{T.T} \; \rightarrow \; \texttt{T} \; \boxed{2} \; \texttt{Int} \; \rightarrow \; \texttt{Int}) \; \cup \; \{\texttt{Bool} \ge \texttt{V}_r\}$	By C1
$((\texttt{T} \; \rightarrow \; \texttt{T}) \; \boxed{2} \; (\texttt{Int} \; \rightarrow \; \texttt{Int})) \; \cup \; \{\texttt{Bool} \ge \texttt{V}_r\}$	By C2
$(\texttt{Int} \; \boxed{2} \; \texttt{T}) \; \cup \; (\texttt{T} \; \boxed{2} \; \texttt{Int}) \; \cup \; \{\texttt{Bool} \ge \texttt{V}_r\}$	By C5

valid checks $\texttt{Int} \ge \texttt{Int}$ (which is true), so the answer is

$$(\{\texttt{Bool} \ge \texttt{V}_r, \texttt{T} \ge \texttt{Int}, \; \texttt{Int} \ge \texttt{T}\}, \; \texttt{V}_r)$$

Which means that $\texttt{f} \; \texttt{x}$ has type $\texttt{V}_r$, provided that $\texttt{Bool} \ge \texttt{V}_r$.

Suppose that the argument and parameter types had been the other way round. The initial tests would follow the same course, but

$$(\forall \texttt{T.T} \; \rightarrow \; \texttt{T}) \; \rightarrow \; \texttt{Bool} \; \boxed{2} \; (\texttt{Int} \; \rightarrow \; \texttt{Int}) \; \rightarrow \; \texttt{V}_r$$

would reduce like this:

$(\texttt{Int} \; \rightarrow \; \texttt{Int} \; \boxed{2} \; \forall \texttt{T.T} \; \rightarrow \; \texttt{T}) \; \cup \; \texttt{Bool} \; \boxed{2} \; \texttt{V}_r$	By C5
$(\texttt{Int} \; \rightarrow \; \texttt{Int} \; \boxed{2} \; \; \forall \texttt{T.T} \; \rightarrow \; \texttt{T}) \; \cup \; \{\texttt{Bool} \ge \texttt{V}_r\}$	By C1
$(\texttt{Int} \; \rightarrow \; \texttt{Int} \; \boxed{2} \; \texttt{T} \; \rightarrow \; \texttt{T}) \; \cup \; \{\texttt{Bool} \ge \texttt{V}_r, \; \texttt{T} \; \text{Fixed}\}$	By C4
$(\texttt{T} \; \boxed{2} \; \texttt{Int}) \; \cup \; (\texttt{Int} \; \boxed{2} \; \texttt{T}) \; \cup \; \{\texttt{Bool} \ge \texttt{V}_r, \; \texttt{T} \; \text{Fixed}\}$	By C5
$\{\texttt{Bool} \ge \texttt{V}_r, \; \texttt{T} \; \text{Fixed}, \; \texttt{Int} \ge \texttt{T}, \; \texttt{T} \ge \texttt{Int}\}$	

but this time *valid* would produce Fail (By S5).

5.6. Recursive generators

In the absence of recursive types it is clear that rules C1-5 will produce a finite set of assumptions, since each rule involves a reduction in size of the comparands. Recursive types have the effect that types are no longer finite, and so recursion on their structure may not terminate. Note that recursive types can arise from expressions that are otherwise not recursive. If y has type $\forall \texttt{V.} \; (\texttt{V} \; \rightarrow \; \texttt{V}) \; \rightarrow \; \texttt{V}$ and f has type $\forall \texttt{T.T} \; \rightarrow \; \texttt{Int} \; \texttt{x} \; \texttt{T}$, then y f will have the recursive type $\texttt{Int} \; \texttt{x} \; \texttt{Int} \; \texttt{x}...$, i.e. $\mu \texttt{V.Int} \; \texttt{x} \; \texttt{V}$ (the type checker will express this as $\texttt{V}$ with $\{\texttt{Int} \; \texttt{x} \; \texttt{T} \ge \texttt{V}, \; \texttt{V} \ge \texttt{T}\}$). If one wanted to use the above algorithm in a system without recursive types it would be necessary to perform an 'occurs check' to ensure that no type variable appeared in its own bounds. For the full system it simplifies the computation of the *closure* operation if such recursive conditions are replaced with conditions containing explicitly recursive types.

The restrictions Ponder imposes on recursive generators ensure that the expansions of recursive calls to type generators are equivalent to the original application; so the application of G marked with $\Uparrow$ will be equivalent to the initial application of G.

$$\texttt{G}[\texttt{V}_1, \dots, \texttt{V}_n] \; \triangleq \; \dots \; \underset{\Uparrow}{\texttt{G}[\texttt{V}_1, \dots, \texttt{V}_n]} \; \dots$$

Since the texts of recursive generators are finite, it is clear that any infinite series of comparisons that could arise must be a cycle. Hence it suffices to introduce a memory into the algorithm, and not perform any comparison ($\geq$ or $\leq$) twice (the assumptions generated must be the same as some that have occurred already).

Note that the same mechanism is necessary for *closure*, in order to avoid checking the same thing repeatedly.

5.7. Optimisations

The description of the algorithm above is simplified. In fact, it is desirable that a type error should be detected as soon as possible (the overloading mechanism of Ponder relies on type-checks that fail).

A number of optimisations to the algorithm are therefore necessary. Instead of computing the sets of assumptions and then checking for consistency at the end it is useful to pass closure of the set as computed so far to succeeding comparisons, and use a special function to insert new elements in the set. This means that inconsistencies of the form $\{V \geq T, V \text{ Fixed}\}$ are discovered immediately, and the comparison may stop. To speed the process of insertion, it is useful to sort the assumptions by variable, and to duplicate cases where two variables are compared, so that the relationship is keyed by both variables.

A further optimisation arises from the observation that it is not necessary to perform an expansion when comparing type generators that are the same. A more efficient approach is to compare only the arguments of the two generators. However, it is necessary to pre-calculate the directions that comparisons will take. For example if $F[A,B] \triangleq A \rightarrow B$ then for $F[A_1,B_1] \geq F[A_2,B_2]$ we need $B_1 \geq B_2$, but $A_2 \geq A_1$. In a comparison $G[T_{1L}, \dots T_{nL}] \geq G[T_{1R}, \dots , T_{nR}]$, each T_{iL} will be compared with the corresponding parameter T_{iR}, either by $T_{iL} \geq T_{iR}$, or by $T_{iR} \geq T_{iL}$ or both. It is a simple matter to pass over the definition of a generator and classify each parameter according to which way it will be compared.

Finally, it makes good sense to remember the assumptions associated with the types of variables in the environment, rather than to carry all the assumptions as one enormous set. Even so, the sets can be quite large and it helps if pairs of inequations like $V_1 \geq V_2$, $V_2 \geq V_1$ are remembered separately as $V_1 = V_2$. Even with these optimisations the algorithm implemented as a Ponder program is lamentably slow.

6. REPRESENTING OBJECTS

Because there are no built-in functions or data-types, objects in Ponder are represented as functional data-structures. In this section I give the definitions of a few types that one would normally expect to be built in.

None of the examples is particularly complex; I merely wish to show that the type system is strong enough for them, and add that it is possible for the compiler to detect the general cases of which these are particular examples, and hence to implement them efficiently.

6.1. Booleans

$$\text{Type Bool} \triangleq \forall T.T \rightarrow T \rightarrow T$$

In which `true` $\triangleq$ $\lambda x.\lambda y.x$ and `false` $\triangleq$ $\lambda x.\lambda y.y$. Any (terminating) function of this type must be equivalent to either `true` or `false`. The Ponder compiler takes advantage of this fact when generating code, so that any application of an object of that type to two arguments is generated as a test and jump, and so is as efficient as if `Bool` had been built in. In fact, the compiler recognises the general case of types of the form $\forall T.T \rightarrow T \rightarrow ... \rightarrow T$, and generates n-way branches as appropriate.

6.2. Pairs

$$\text{Type Pair}[L,R] \triangleq \forall Res.(L \rightarrow R \rightarrow Res) \rightarrow Res$$

In which pairs are represented as functions that may be applied to $\lambda x.\lambda y.x$ or $\lambda x.\lambda y.y$ to return their first or second component respectively. Thus the λ-expression for the pair constructing function is $\lambda l.\lambda r.\lambda u.u \ l \ r$ and the functions to take the left and right elements of a pair are $\lambda p.p \ true$ and $\lambda p.p \ false$ respectively.

6.3. Options

Options are things that may or may not be there, like the tail of a list or the daughters of a node in a tree. An option is essentially a special case of a union (see below) in which the second type is empty.

$$\text{Type Option}[T] \triangleq \forall R.(T \rightarrow R) \rightarrow R \rightarrow R$$

i.e. options are functions that either apply a function to the thing they hold, or return another result. In other words, if *opt* is an optional integer, then it will either be $\lambda f.\lambda x. f$ n for some integer n, or it will be $\lambda f.\lambda x.x$ (which is called *nil*), so

$$opt (\ \lambda x.2 \times x) \ 0$$

will either return twice n if it is there, or zero if it isn't.

6.4. Lists

The previous two definitions allow us to define lists:

$$\text{Rectype List}[T] \triangleq \text{Option}[\text{Pair}[T, \text{List}[T]]]$$

Which means that a list will either be *nil* (as in `Option`'s above), or will accept a function to which it will pass the head and tail of the list.

6.5. Binary Unions

Unions are like options, except that the thing is either one type of thing or another type of thing:

$$\texttt{Type Union[T}_1\texttt{,T}_2\texttt{]} \triangleq \forall \texttt{R(T}_1 \rightarrow \texttt{R)} \rightarrow \texttt{(T}_2 \rightarrow \texttt{R)} \rightarrow \texttt{R}$$

so if u is of type `Union[A,B]` then u $\texttt{f}_a$ $\texttt{f}_b$ applies either $\texttt{f}_a$ if the object in the union is of type `A`, or $\texttt{f}_b$ if it is of type `B`.

Although these are binary unions, the use of the syntactic definition facilities of Ponder makes it possible to use them quite comfortably for any number of types.

6.6. Abstract types

In their paper [MP85], Plotkin and Mitchell describe a means of representing abstract data types in terms of *existentially* quantified types. At first sight one might assume that this technique cannot be used in Ponder because of the absence of existential quantifiers. Fortunately the well known correspondence in logic between universal and existential quantifiers applies to types as well. That is, where one wants an object of type $\exists \texttt{T.G[T]}$ one can instead use an object of type $\forall \texttt{V.} (\forall \texttt{T.G[T]} \rightarrow \texttt{V}) \rightarrow \texttt{V}$. Here the concrete properties of an abstract object x are hidden inside the function $\lambda \texttt{f.f}$ x, by the requirement that f be polymorphic. This gives a 'classical' version of $\exists$, to which should be contrasted the constructive version of Cardelli and MacQueen [CM85].

Note that in Milner's system abstract type definitions of `Bool`, `Pair`, `Option` and `Union` would not be allowed, and the use of existential types in the manner indicated above is precluded by the exclusion of polymorphic parameters caused by the absence of local quantification.

7. INADEQUACIES

Ponder has been used to write a large number of small programs (an interactive calculator for example), and some larger programs (such as the spreadsheet). Unfortunately a number of things are lacking and are difficult or impossible to provide without altering the language.

7.1. Dependent types etc

One assumption that was taken as fundamental in the design of the type system was that all type information could be checked and removed at compile-time. Although I still support this as a general principle, the discussions at the Appin workshop made it clear that there are problems associated with static typing in a persistent environment. Atkinson and Morrison [AM85] describe some of these problems and propose solutions. The expressive power of static typing has not yet been exhausted, however, and I feel that many of the problems may be solved by coding type information in some way at run time. Examples of this are the union types described above, and the dependent types found in constructive type theory.

8. CONCLUSION

The type checker described above differs slightly from the one distributed with the current Ponder system. This previous version has several infelicities that cause it to reject some expressions that are type-valid. This was caused by an attempt to make the type-checker 'guess' a particular type for every sub-expression, something which although impossible to do accurately is desirable for helpful error messages.

Nonetheless, users of the Ponder compiler find the type-checking system sufficiently powerful for most of their requirements, and have been using the previous (pessimistic) type-checker for more than four years with little complaint.

9. ACKNOWLEDGEMENTS

I am grateful to Mike Gordon for patient supervision, and to Dave MacQueen for some useful conversations that helped to clarify my thoughts about the algorithms. Further thanks are due to Alan Mycroft who discovered some mistakes in a previous version of the algorithm.

The work has been funded by the Science and Engineering Research Council of Great Britain.

REFERENCES

[BMS80] [CM85] [DD80a] [FAIR82] [GMW79] [MATT85b] [MILN78]
[MP85] [MS82] [MS84] [REYN86] [ROBI65] [VANW75]

Part II

Database Types in Programming Languages

Chapter 7

Data Types For Database Programming

Peter Buneman[1]
University of Pennsylvania

1. INTRODUCTION

There are three data types that find ubiquitous use in data base programming. They are

- record types,
- index or table types, and
- set or relation types.

It is suggested in this paper that the three types can be derived from one underlying construct, a *partial function*, or *map* as we shall call it that is well-behaved with respect to inheritance. Moreover the behaviour of these functions provides a link with the substantial research on database theory [ULLM82] which, sadly, has yet to have any great influence on the types of database programming languages although it is generally regarded as extremely important in database design. Another reason for wanting to find a unifying framework for these types is to form a better understanding of how the *functional* data model [SHIP81] [BUNE82] relates to the *relational* data model [CODD70] and how these in turn fit with inheritance.

To take an example of how these various types are used in database programming languages, one of the earliest such languages is Pascal-R [SCHM77], which extends Pascal with a **relation** data type so that one can declare for example,

```
type Employee = record
                    EmpNo : Integer;
                    EmpName : array[1..30] of char;
                       . . .
                end
var EmpRel : relation <EmpNo> of Employee;
 . . .
```

There are various methods in the language for performing the usual operations of the relational algebra; in addition, it is possible to use a relation such as `EmpRel` as an index that, given a key of type **integer**, returns a record of type `Employee`. Thus `EmpRel[1234]` returns that record - if it exists - whose `EmpNo` is 1234. In some sense the relation `EmpRel` is doing double duty both as a set and as an index. More recent database programming languages such as Taxis [MBW80], Adaplex [CCA83] and Galileo [ACO83] provide similar constructs although in some cases the index type is not directly available to the programmer but is nevertheless implemented as a method for optimising the manipulation of relations or similar types.

---

1 This research was carried out on a British Science and Engineering Council Fellowship at the University of Glasgow

Although these types are all implemented, they are not necessarily treated unifomly within the language. One reason is undoubtedly that two of these types, index and relation, may be singled out as *persistent* types and the implementation of persistent types may restrict what can, for example, serve as a component type for a relation or the input and output types for an index. A survey of the problems associated with persistence and data types is to be found in [BA85]. A more fundamental problem is what the precise definition of these types is, and how they interact with one another. In particular, one might ask whether there is a single underlying type of which each of these three types is a manifestation. Another question is how the basic operations of, say, the relational algebra act on these types. For example, is the most "natural" of all relational operations, the *natural join*, a primitive operation, is it something that the programmer can define in terms of more primitive operations., or is it something that is not directly available? In the case that it is primitive or programmer-defined, what is its type?

This paper describes a simple notation for partial functions, and an informal description of inheritance. A more formal description of *maps* - partial functions that are well-behaved with respect to inheritance is then derived, and it is shown how records, relations and indexes can all be derived from this one construct. It is also indicated, briefly, how this notion of a partial function connects with relational database theory. A final section outlines how type-checking can be implemented to give a somewhat more faithful representation of relational and other database operations in a typed programming language.

2. RECORDS AND INHERITANCE

The clue to providing a unified set of types for databases comes from the idea of *inheritance*. The importance of inheritance (sometimes called *generalisation* or *subsumption* has been recognised for some time in programming languages [XERO81], databases [SS77] and in semantic networks. Recent work [CARD84] has shown that, at the level of types, inheritance can be cleanly combined with functional programming; and [AITK84] shows that inheritance can itself serve to model computation and to provide a form of logic programming. Here we shall initially describe inheritance at the level of values, as is done, for example, in [AS81]. To provide an informal introduction some notation is introduced for describing partial functions on a finite set of values. The expression,

$$\{'Susan' \Rightarrow 3490; 'Peter' \Rightarrow 7731; 'Karen' \Rightarrow 8535\}$$

describes a very small telephone directory - a partial function from strings to integers. If the range of the partial function contains just one element, { }, we may use an abbreviated description

$$\{7; 123; 22\}$$

for

$$\{7 \Rightarrow \{\}; 123 \Rightarrow \{\}; 22 \Rightarrow \{\}\}$$

Such partial functions apparently correspond to subsets of the input set of values, but we shall shortly modify our definition of partial function so that this correspondence is not entirely correct. When the input to the partial function is a set of labels as in

$$\{Idno \Rightarrow 12345; Name \Rightarrow 'Jones'; BirthYear \Rightarrow 1957\}$$

we may think of the partial function as a record whose output is taken from some heterogeneous space of values.

Now there is a natural ordering on partial functions. If we think of records as descriptions of some "real world" object or event, this ordering corresponds to the notion of "better description". For example

$$\{IdNo \Rightarrow 1234;\ Name \Rightarrow \text{'}Jones\text{'};\ BirthYear \Rightarrow 1957\}$$

is a better description than

$$\{IdNo \Rightarrow 1234;\ Name \Rightarrow \text{'}Jones\text{'}\}$$

by virtue of having more fields and agreeing on those fields that are common to the two descriptions. Similarly

$$\{IdNo \Rightarrow 12345;\ Address \Rightarrow \{City \Rightarrow \text{'}Philadelphia\text{'};\ Zip \Rightarrow 19118\}\}$$

is a better description than

$$\{IdNo \Rightarrow 12345;\ Address \Rightarrow \{City \Rightarrow \text{'}Philadelphia\text{'}\}\}$$

because the field values in one are themselves descriptions and are better defined in one than the other.

This last expression has introduced the possibility that partial functions may be "higher order" and that the input and output functions may be spaces on which there is an ordering. The question now arises whether any partial function we can write down in this notation is allowable. Consider the following three expressions.

$$\{\{Emp\# \Rightarrow 1234\} \qquad \Rightarrow \{Name \Rightarrow \text{'}J.\ Brown\text{'};\ Office \Rightarrow Philadelphia\};$$
$$\{Emp\# \Rightarrow 1234;\ ShoeSize \Rightarrow 10\} \Rightarrow \{Name \Rightarrow \text{'}K.\ Smith\text{'}\}\}$$

(a)

$$\{\{Stud\# \Rightarrow 3456\} \qquad \Rightarrow \quad \{Name \Rightarrow \text{'}D.\ Dare\text{'}\};$$
$$\{Course\# \Rightarrow \text{'}CIS123\text{'}\} \qquad \Rightarrow \quad \{CName \Rightarrow \text{'}Database\ Systems\text{'}\}$$
$$\{Stud\# \Rightarrow 3456;\ Course\# \Rightarrow \text{'}CIS123\text{'}\} \qquad \Rightarrow \quad \{Name \Rightarrow \text{'}D.\ Dare\text{'}; \qquad \text{(b)}$$
$$CName \Rightarrow \text{'}Database\ Systems\text{'};$$
$$Grade \Rightarrow \text{'}A\text{'}\}\}$$

$$\{\{Emp\# \Rightarrow 1234\} \qquad \Rightarrow \quad \{Name \Rightarrow \text{'}J.\ Brown\text{'};\ Office \Rightarrow Philadelphia\}$$
$$\{Emp\# \Rightarrow 1234;\ ShoeSize \Rightarrow 10\} \Rightarrow \quad \{Name \Rightarrow \text{'}J.\ Brown\text{'};\ Office \Rightarrow Philadelphia\}$$
$$\{Emp\# \Rightarrow 1234\} \qquad \Rightarrow \quad \{Name \Rightarrow \text{'}J.\ Brown\text{'}\}\}$$

(c)

Example (a) is badly behaved. In return for a better input it has produced a less informative - and contradictory - output. Example (b) is the sort of behaviour one might expect from a database system. There is extra information to be gained by providing a better specified input. Example (c) is redundant in that we can infer the second and third input-output pairs from the first, but we can nevertheless consider these pairs as part of the partial function. Of course, it is difficult to imagine how a typed programming language would allow us to assign a type to any of these examples, although several database management systems allow us to represent data like those of example (b).

Example (c) poses a question of representation. Should we think of the "redundant" part of the partial function as not being part of the partial function at all, or should we adopt the notion that a partial function is really a (binary) relation that includes all the redundant parts. We adopt the latter notion.

3. A DOMAIN THEORETIC DESCRIPTION OF PARTIAL FUNCTIONS

In order to prevent badly behaved partial functions such as the one we have just seen, we need to set up some formal apparatus to describe which partial functions are admissible with respect to inheritance. The mathematical results in this section are all well-known in domain theory; only their interpretation with respect to databases is new. We shall define a partial function as a subset F of $V_1 \times V_2$ where V_1 and V_2 are domains. The conditions we impose are

1. $(\perp_1, \perp_2) \in F$.
2. $(x,y) \in F$, $x' \sqsupseteq x$, and $y \sqsupseteq y'$ imply $(x',y') \in F$
3. $(x,y_1) \in F$ and $(x,y_2) \in F$ imply $(x, y_1 \sqcup y_2) \in F$

The second of these conditions guarantees that the partial function is well-behaved with respect to inheritance; the third ensures that the relation F is "functional", namely that there is a unique "best" output for a given input. The first condition is for convenience in simplifying further definitions. I shall use the term *map* to describe a subset of $V_1 \times V_2$ with these properties and use $V_1 \mapsto V_2$ to denote the set of such maps. Programming language semanticists will recognise the definition of a map to be similar to the definition of an approximable mapping in [SCOT82], and this connection deserves much closer examination. However, our purpose in this paper is to examine the consequences of regarding a map as a practical data constructor.

Maps are ordered by inclusion. If F_1 and F_2 are maps in $V_1 \mapsto V_2$, we can define

$$F_1 \sqcap F_2 = F_1 \cap F_2$$

and

$$F_1 \sqcup F_2 = \cap \{F \mid F \in V_1 \mapsto V_2, F \supseteq F_1 \text{ and } F \supseteq F_2 \}$$

the latter only being defined when the intersection is non-empty. Thus maps themselves form a domain.

We can call a map *elementary* if it is generated by a single pair of elements, i.e. it is of the form

$$\{(z,\perp_2) \mid z \in V_1\} \cup \{(x',y') \mid x' \in V_1, y' \in V_2, x' \sqsupseteq x, y \sqsupseteq y'\}$$

for any $x \in V_1$ and $y \in V_2$. A map is defined to be *finite* if it is of the form $F_1 \sqcup F_2 \sqcup \ldots \sqcup F_n$ where $F_1, F_2, \ldots, F_n$ are elementary. The notation of the previous section provides us with a method for describing finite maps (provided the notation is consistent with a finite map.) It follows from these definitions that if F_1 and F_2 are finite then so are

$F_1 \sqcup F_2$ and $F_1 \sqcap F_2$ whenever these exist. Another useful observation is that the composition of two maps,

$$F \circ G = \{(x,z) \mid \exists \, y. \, (x,y) \in F, \, (y,z) \in G\}$$

is a map and that $F \circ G$ is finite if F is finite or if G is finite.

Having covered this groundwork, let us look at some simple classes of maps.

1. Suppose V is a flat domain of values (e.g. the integers) and TRIV is the trivial domain containing one (non-bottom) element $\{\}$. The maps in $V \mapsto$ TRIV correspond to subsets of $V \setminus \{\perp\}$ with $\sqcap$ and $\sqcup$ respectively corresponding to set intersection and set union. The ordering is the containment ordering.

2. Suppose Λ is a finite flat domain of labels and V is any domain. The maps in $\Lambda \mapsto$ V are the *records* over V with the ordering described informally in the previous section. We shall call this domain $\mathcal{R}(V)$

3. Now consider the finite maps in $\mathcal{R}(V) \mapsto$ TRIV. It is claimed that we can identify these with the relations over V with $\sqcap$ denoting the *natural join*.

The last example requires some elaboration. In the first place, since $\mathcal{R}(V)$ is not a flat domain there is a distinction between relations and sets (as defined in case 1 above). In fact, the members of $\mathcal{R}(V) \mapsto$ TRIV are in 1-1 correspondence with the upward closed subsets, or *filters* of $\mathcal{R}(V)$. Secondly, in database programming languages, we usually think of relations as a set of uniformly typed, flat, records (i.e. records whose values lie in a flat domain.) However there is no need for this in defining the natural join. For example, the natural join of

```
{  {Name ⇒ 'J. Doe';   Dpt ⇒ 'Sales';  Add ⇒ {Cty ⇒ 'Moose'            } };
   {Name ⇒ 'M. Mac';   Dpt  ⇒ 'Manuf';
   };
   {Name ⇒ 'N. Bug';                     Add ⇒ {            State ⇒ 'Mo'}}}
```

and

```
    {  {Dpt ⇒ 'Sales';   Add ⇒ {                    State ⇒ 'Wy' }
       };
       {Dpt ⇒ 'Rsrch';   Add ⇒ {Cty ⇒ 'Billings'                 }
       };
       {Dpt ⇒ 'Manuf';  Add ⇒ {            State ⇒ 'Mo' }}
       }
```

is

```
{{Name ⇒ 'J. Doe';                Dpt ⇒ 'Sales';        Add ⇒ {Cty ⇒
'Moose';          State ⇒ 'Wy'    }                     };
  {Name ⇒ 'M. Mac';  Dpt ⇒ 'Manuf'; Add ⇒ {             State ⇒ 'Mo' }
  };
  {Name ⇒ 'N. Bug';  Dpt ⇒ 'Manuf'; Add ⇒ {             State ⇒ 'Mo' }
  };
  {Name ⇒ 'N. Bug';  Dpt ⇒ 'Rsrch'; Add ⇒ {Cty ⇒ 'Billings';          State
⇒ 'Mo'              }}}}
```

Thus the definition of natural join extends naturally to "ragged" and "non-flat" (non first-normal-form) relations. In the case that R1 and R2 are each uniformly typed, our definition agrees with the usual definition of natural join, and if R1 and R2 have the same type the natural join, as expected, defines the intersection, and R1 $\sqcup$ R2 gives the union.

We can take this domain-theoretic approach to relations further, and from a relation R defined on $\mathcal{R}(V) \mapsto$ TRIV, define a map R' in $\mathcal{R}(V) \mapsto \mathcal{R}(V)$ by

$$(x,y) \in R' \text{ iff } \forall r.(r,\{\}) \in R \text{ and } r \sqsupseteq x \text{ imply } r \sqsupseteq y$$

R' belongs to the special class of maps known as *closures*. One way of characterising a closure is by the following rules. A map F is a closure if

1. $x \sqsubseteq y$ implies $(x,y) \in F$

2. $(x,y) \in F$ implies $(x \sqcup w, y \sqcup w) \in F$

3. $(x,y) \in F$ and $(y,z) \in F$ imply $(x,z) \in F$

The reason for casting the closure property in this form is that any database theorist should immediately recognise this definition as Armstrong's axioms for *functional dependencies* (replace $\sqsubseteq$ by $\supseteq$ and $\sqcup$ by $\cup$ on the set of labels.) The precise connection is beyond the scope of this paper, however it is worth mentioning that we can derive several of the basic results in relational database theory by this domain-theoretic approach and extend them to ragged, non-flat relations. It should also be pointed out that the derivation of closure given above is *not* the way one actually obtains Armstrong's axioms. Moreover, for various technical reasons it is necessary to take the elements of a relation R, the set $\{r_1, r_2,..., r_n\}$ such that R is the union of elementary maps generated by $(r_i,\{\})$ $(1 \leq i \leq n)$, as an *inconsistent* set, i.e. a set of elements which is pairwise inconsistent (any pair does not have a defined $\sqcup$.)

4. IMPLICATIONS FOR DATABASE TYPE SYSTEMS

The previous section provided some theoretical justification for using maps as a basic data constructor for database programming languages since we could use it to represent (and perhaps implement) record, relation and index types. The purpose of this section is to suggest how such structures might be practically incorporated into a typed programming language. However, we apparently achieved this uniformity by treating labels as *values* and this may give us rather more than we had bargained for. For example it is difficult to attach a meaning to maps such as

$$\{Name \Rightarrow 'Jones'; 4 \Rightarrow 'emu'; 17 \Rightarrow Age\}$$

since they appear to be ill typed (although they are perfectly reasonable expressions.) What kind of type system will exclude such expressions? Fortunately [CARD84] has provided us with most of the groundwork, and his type system for Amber with minor modifications will support precisely the kinds of maps in which we are interested. What follows is a very brief account of the modifications to this system that will support map types. The reader is referred to [CARD84] for further details.

To start with, we remove record types from the domain of values, but extend the domain to include maps and labels[2]; thus we express the domain of values as

$$\begin{aligned} \mathbf{V} &= \mathbf{B} + \mathbf{L} + \mathbf{M} + \mathbf{F} + ... \\ \mathbf{M} &= \mathbf{V} \mapsto \mathbf{V} \\ \mathbf{F} &= \mathbf{V} \rightarrow \mathbf{V} \end{aligned}$$

where $\mathbf{B}$ is a flat domain (or sum of flat domains) of basic values (such as integers, strings, booleans), $\mathbf{L}$ is a flat domain of labels, $\mathbf{M}$ is the domain of finite maps, $\mathbf{F}$ is the domain of functions etc.

Now there is a slight problem in doing this. In order to define an ordering on maps, we needed (at some level) to have equality defined on the domain of values $\mathbf{V}$, and equality is not decidable when $\mathbf{F}$ contains elements corresponding to arbitrary expressions in the λ-calculus. However, higher-order languages (languages in which functions are values) such as ML [MILN83] and PS-algol [AM85] get round this problem either by having equality that *fails* on functions, or by using some restricted notion of equality such as textual equality or referential equality. We shall assume this has been done.

We now need to assign types to labels and maps. Since labels are values, we must severely restrict the kinds of "run-time" computation we can do if we want to preserve the kinds of type-checking common in languages with records. To do this we assign each label its own type. Moreover, since context should make clear whether we are referring to values or types, we shall use the same term to denote both the value and the type. Thus the type of the label IdNo is IdNo. We now want to have rules that will type an expression such as

$$\{Idno \Rightarrow 12345; \; Name \Rightarrow 'Jones'; \; BirthYear \Rightarrow 1957\}$$

as

$$\{Idno \Rightarrow int; \; Name \Rightarrow string; \; BirthYear \Rightarrow int\}$$

We also want to allow other kinds of maps in which the domain and range are of uniform type. To do this we add a type expression $\tau_1 \mapsto \tau_2$ where τ_1 and τ_2 are type expressions. Thus the type of

$$\{'Susan' \Rightarrow 3490; \; 'Peter' \Rightarrow 7731; \; 'Karen' \Rightarrow 8535\}$$

should be *string* $\mapsto$ *int*.

To do this we use the following typing rules for maps

1.   if $e_1 : \tau, e_2 : \tau,...,e_n : \tau$ and $e'_1 : \tau', e'_2 : \tau',...,e'_n : \tau'$ then
   $\{e_1 \Rightarrow e'_1; e_2 \Rightarrow e'_2;...; e_n \Rightarrow e'_n\} : \tau \mapsto \tau'$

2.   if $l_1, l_2,...,_n$ are labels and $e'_1 : \tau'_1, e'_2 : \tau'_2,...,e'_n : \tau'_n$
   then $\{l_1 \Rightarrow e'_1; l_2 \Rightarrow e'_2;...;l_n \Rightarrow e'_n\} : \{l_1 \Rightarrow \tau'_1; l_2 \Rightarrow \tau'_2;...;l_n \Rightarrow \tau'_n\}$

3.   if $e : \sigma \mapsto \tau$ and $f : \sigma' \mapsto \tau'$
   then $(e \sqcap f) : (\sigma \sqcup \sigma') \mapsto (\tau \sqcap \tau')$
   and $(e \sqcup f) : (\sigma \sqcap \sigma') \mapsto (\tau \sqcup \tau')$
   provided all meets and joins are defined.

---

[2] variant records or sums are not considered here.

Of these rules (1) follows Cardelli's type rules for records, (3) his rules for functions, which behave very much like maps. Only (2) is at all new.

Some examples may clarify the use of these rules. In the first place, none of these rules (and hopefully no others) will assign a type to expressions like

$$\{Name \Rightarrow \text{'Jones'}; 4 \Rightarrow \text{'emu'}; 17 \Rightarrow Age\}$$

Second, we can examine the type of natural join and other relational operations. Consider the declarations

type *Person* = {*IdNo* ⇒ *int*; *Name* ⇒ *String*; *DeptNo* ⇒ *int*}
type *Department* = {*DeptNo* ⇒ *int*; *DName* ⇒ *string*};

val *PersonRel* : *Person* ↦ {} = ...;
val *DepartmentRel* : *Department* ↦ {} = ...;

that describe the relations *PersonRel* and *DepartmentRel*.

The natural join *PersonRel* ⊓ *DepartmentRel* of these two relations has type, inferred by rule 3, (*Person* ⊔ *Department*) ↦ {} or

$$\{IdNo \Rightarrow int; Name \Rightarrow string; DeptNo \Rightarrow int; Dname \Rightarrow string\} \mapsto \{\}$$

which accords with our intuition for the type of a natural join. Note that since ⊔ is not always defined, the inference rules may fail to find a type for the natural join. This would happen if, for example, we rewrote the type of *Department* as {*DeptNo* ⇒ *string*; *DName* ⇒ *string*}.

Third, the rules do assign a type to expressions such as

$$\{EmpNo \Rightarrow IdNo; LastName \Rightarrow Name\},$$

a map in which both the input and output are labels. Such maps are extremely useful in database work for relabelling records. Relabelling of a record is nothing more than composition of maps, thus

$$\{EmpNo \Rightarrow IdNo; LastName \Rightarrow Name\} \circ \{IdNo \Rightarrow 12345; Name \Rightarrow \text{'Jones'}\}$$

evaluates to

$$\{EmpNo \Rightarrow 12345; LastName \Rightarrow \text{'Jones'}\}$$

and is again well-typed. Note that {*EmpNo* ⇒ *IdNo*; *LastName* ⇒ *Name*} denotes a type as well as a value. A special case of relabelling is *projection*, which is extended to relations in the relational algebra. For example

$$\{IdNo \Rightarrow IdNo\} \circ \{IdNo \Rightarrow 12345; Name \Rightarrow \text{'Jones'}\}$$

"projects" onto *IdNo* to produce {*IdNo* ⇒ 12345}.

The idea that labels are types in the sense described is somewhat contentious, but may be needed if we are to write suitably generic functions for database work. Consider the problem of writing a *transitive closure* function for a relation. There is no difficulty in expresssing the transitive closure in Prolog or any other language that treats parameters positionally, but what can we do using the relational operators of a database programming language? Specifically, suppose we are given a relation of type $\{c1 \Rightarrow int; c2 \Rightarrow int\} \mapsto \{\}$ and we want the output to be a relation of the same type. This means that we have to do successive joins of a relation in which we identify the $c1$ label (column) of one relation with the $c2$ field in the other. Thus the join is not a natural join. In most programming languages that attempt to represent relations, this has to be done by unpacking the relations, forming the identification among records, and reconstructing the result from tuples. Here is a function that performs the task using natural joins and relabelling.

> **val** $TC(R : \{c1 \Rightarrow int; c2 \Rightarrow int\} \mapsto \{\} = TC1(R, NUL, R)$
> **where**
> **val rec** $TC1(R,S,T : \{c1 \Rightarrow int; c2 \Rightarrow int\} \mapsto \{\}) =$
>   **if** $T \sqsubseteq S$ **then** S
>   **else** $TC1(R, S \sqcup T,(\{c1 \Rightarrow c1; l \Rightarrow c2\} \circ \circ R) \sqcap (\{l \Rightarrow c1; c2 \Rightarrow c2\} \circ \circ R)$

In order to perform the relabelling for the join, we have introduced a new label l. NUL is the empty relation and $\circ \circ$ is the extension of relabelling for a relation. The point of introducing this example is that we would like to be able to write a *generic* transitive closure function. To do this some form of label parameterization is essential.

5. FURTHER RESEARCH

We have shown that taking maps as data constructors and treating labels as a special class of values provides a uniform treatment of data types for database programming. What we have yet to do is to present an adequate set of operators for maps. We have seen that meet ($\sqcap$), join ($\sqcup$) and composition ($\circ$) are useful and allow us to produce most of the operations of the relational algebra. The adequacy of these operations depends, of course, on what is available in the rest of the language. Database programming languages usually allow some form of iteration over relations (or some similar data type) and one can therefore implement any operation by decomposing and reconstructing relations a record at a time. The question of what is adequate therefore is not one of computational power, but what is most convenient. This is something that can only be answered by substantial experiments with various combinations of operators.

Another omission is a discussion of implementation. If we use type constraints, as has been suggested, to ensure that all maps are uniformly typed, then a B-tree or hash-table mechanism is sufficient. In fact it is interesting to note that PS-algol uses precisely the same mechanism to resolve record labels at compile time as it does to implement run-time tables. However, were we to extend our hypothetical language to deal with ragged relations (relations with nulls), we would be faced with a more complicated indexing problem. In this connection some of the techniques suggested for implementing universal relations might be appropriate.

A more difficult problem though is to examine type parameterised programming. Languages such as Russell [DD80] and Poly allow types to be treated as objects so that functions can be constructed that take types as parameters and produce new types as results. We have suggested that labels are values with individual associated types. Thus there is a special class of label types and one might use this in a type- and label-parameterised version of the transitive closure operation shown above. Perhaps the most interesting prospect of treating labels as parameters is that database schemas and semantic networks are large labelled graphs. A database management system takes such a graph, checks it for consistency, and produces what is in effect an abstract data type. It may be that this approach to data types will allow us to treat database systems as large parameterised data types and remove once and for all the lacuna between database systems and programming languages.

REFERENCES

[ACO83] [AITK84] [AM85] [AS81] [BA85] [BUNE82] [CARD84] [CCA83] [CODD70] [DD80] [MBW80] [MILN83] [SCHM77] [SCOT82] [SHIP81] [SS77] [ULLM82] [XERO81]

Chapter 8

The Type System of Galileo

A.Albano
F. Giannotti[1]
R.Orsini
D. Pedreschi
Università di Pisa

ABSTRACT *Galileo is a conceptual language, i.e., a programming language for database applications which supports both the abstraction mechanisms of modern programming languages (data types, abstract data types, modules) and the abstraction mechanisms of semantic data models (SDM) (classification, aggregation, generalization). Unlike other conceptual languages, Galileo is a strongly typed language which exploits the benefits of data types for modeling the abstract knowledge of information systems. The type system of Galileo has been designed to deal with the intensional aspects of the SDM abstraction mechanisms, while the class and subclass mechanism is provided to deal with the extensional aspects. The following relevant features of the type system will be emphasized: type hierarchies and declarative overloading, which provide a kind of polymorphism in a strongly typed language whose range of applications goes beyond the scope of conceptual modeling.*

1 INTRODUCTION

Nowaday there is a growing interest in conceptual languages, i.e. programming languages designed to provide a set of high-level abstraction mechanisms to allow the easy implementation of database applications. In particular, to model the entities of the slice of reality being modeled, their properties and the relationships between them, a semantic data model (SDM) has been proposed with the following abstraction mechanisms [BMS84]:

Classification. The entities sharing common properties are grouped in classes. The names of the classes denote the elements present in the database. The elements of a class are represented uniquely, i.e., only one copy of each element is allowed.

Aggregation. The elements of classes are aggregates, i.e., they are abstractions with heterogeneous components and may have elements of other classes as components. Consequently, relationships among entities are represented by aggregations and not with another mechanism as in an hierarchical or network data model.

---
1   This work was supported in part by Ministero della Pubblica Istruzione.
Present Address: Sipe Optimation, Vicola San Pierino, 4 - 56100 Pisa, Italy.

Generalization. Classes can be organized in a hierarchy through a partial order relationship, often referred to as the ISA hierarchy. Classes related in this way model entities that play different roles in the real world, and may be described by different levels of detail. If A is a subclass of B, the following properties hold: a) in every state, the elements of A are a subset of the elements of B; b) the elements of A inherit all the properties of the elements of B.

An open problem is whether a conceptual language is more effective when based on strong typing. For instance, Galileo is strongly typed [ACO85], Taxis is not [MBW80], while ADAPLEX has been designed starting from ADA, which is a strongly typed language, but features have been included which prevent static type checking of operations [CCA83]. In a strongly typed programming language the type of each identifier, data structure component, expression, function and parameter can be determinable statically, during translation. When a language has been designed with this property in mind, then it is possible to verify statically that an operator is applied to well typed arguments, that a function is invoked with parameters compatible with those declared, that the result of a function is of the type declared, and so on. Although any statically detectable error could also be detected at run-time, it is desirable to do static checks for the following basic reasons: Firstly, programs can be safely executed disregarding any information about types; secondly, the language offers considerable benefits in applications testing and debugging, since the type-checker detects a large class of common programming errors without the need of executing the programs, while error checking at run-time could be detected actually by providing test data that cause the error to be raised.

Of course, a strongly typed language is interesting as long as it also provides expressive abstraction mechanism to construct useful, application oriented types, and so it is not too restrictive in its expressive power. To improve the flexibility of a strongly typed language, different strategies have been followed to include in the type system kinds of polymorphism, i.e. the possibility that some values and variables may have more than one type [CW85]:

- *Parametric polymorphism.* This is the purest form of polymorphism: Type expressions may contain type variables, indicating, for example, that a function can work uniformly on a class of arguments of different, but structurally related types.

- *Type hierarchies.* This is an example of "inclusion polymorphism": If a type *T* is a subtype of the type *T'*, then a a value of *T* can be used as a value of *T'*, but not vice versa [ALBA83] [ACO85] [AOO86] [CARD84].

- *Overloading.* This is a purely syntactic way of using the same name for different operators; the compiler can resolve the ambiguity at compile time, and then proceed as usual.

In designing Galileo the last two strategies were explored to achieve flexibility in a strongly typed conceptual language, since type hierarchies were necessary to support the generalisation mechanism of semantic data models, and overloading in abstract data types definitions was thought to be the simplest way to get a declarative mechanism to describe the structural properties of databases. On the other hand, it is still an open problem how to integrate these features with a parametric polymorphism.

The next section discusses the concrete types and the notion of type hierarchies. Section 3 discusses abstract data types and overloading of operators. Section 4 shows how these mechanisms meet the requirement to model the abstraction mechanisms of semantic data models.

2. CONCRETE TYPES AND TYPE HIERARCHIES

All denotable values of Galileo possess a type. A type is a set of values, together with the primitive operators that can be applied to these values. **num, bool, string** and **null** are predefined types. 3, true and "Galileo" are values of type **num, bool** and **string** respectively. Real numbers, truth-values and strings are equipped with the usual operators. **null** is a set whose only defined element is **nil**, equipped with the equality operator.

A set of *type constructors* exists to define types for structured values: ordered pairs (cartesian product), tuples (unordered labelled cartesian product), variants (discriminated union), sequences, modifiable values and functions. Type constructors provide rules for the definition of *type expressions*. For instance, $\times$ and $\rightarrow$ are the type constructors for the cartesian product and function space, respectively, and so (**num** $\times$ **bool**) is the type for pairs such as "(3, false)", and ((**num** $\times$ **num**) $\rightarrow$ **num**) is the type for functions such as "+".

The concrete type declaration "**type** Id := TypeExp" binds the name Id to the type expression TypeExp, thus allowing to use Id as an abbreviation for TypeExp, as in "**type** BoolPair := **bool** $\times$ **bool**".

The sentence "the expression E has type t" (briefly "E: t") means that the evaluation of E yields a value of type t. The type system of Galileo has been designed to assign a type to every expression by means of textual inspection. This type-checking task is accomplished in a semi-automatic fashion, the user being compelled to provide type information only in a few crucial constructions, such as for formal parameters in function definitions. Indeed, the user can provide explicitly types for expressions with the type-forcing expression "E: t". However, the type-checker is able, in general, to *infer* types for well typed expressions by means of *typing rules,* which describe how to assign a type to a composite expression in terms of the types of its subexpressions. A typing rule has the general form:

$$E_1: t_1, \ldots, E_k:t_k \ , \ Prop$$
$$\overline{}$$
$$C(E_1, \ldots, E_k) : t$$

where Prop is some condition to be satisfied and $C(E_1, \ldots, E_k)$ is an expression built upon the subexpressions E_i's. The rule has to be read: if it is possible to assign type t_i to each E_i and Prop holds, then type t can be assigned to $C(E_1, \ldots, E_k)$. As a first example, the typing rule for pair expressions looks like:

$$E_1 : t, \ E_2 : u$$
$$\overline{}$$
$$(E_1, E_2) : t \times u$$

so that the type **num** $\times$ **bool** can be inferred for the pair "(3, true)".

Since types are interpreted as appropriate subsets of the value domain of the language, it is quite natural to establish a partial ordering among types based on *set inclusion* : this is the key semantic idea underlying type hierarchies. We say that a type t is included in, or is a subtype of, another type u when all the values of t are also values of u, i.e. exactly when t, considered as a set of values, is a subset of u. This general semantic notion of inclusion finds its syntactic counterpart in a set of *compatibility rules* for the type expressions of the language, which allows to check when a type t is a *subtype* of u. In other words, we will define a partial order relation **is** over the type expressions with the property

$$t \textbf{ is } u \implies [u] \supseteq [t]$$

where $[t]$ is the set denoted by type expression t, and "t **is** u" has to be read "t is a subtype of u". Consequently, any value of type t can be safely used where a value of type u is expected:

$$t \textbf{ is } u \implies (V: t \implies V: u) \quad \text{for any value V.}$$

This turns to the very basic typing rule of the system:

$$\frac{E: t, t \textbf{ is } u}{E: u} \tag{1}$$

The definition of the subtype relation involves a general rule ("t **is** t" holds for any type t), and a set of different rules for the different type constructors. For instance, for the cartesian product types the following compatibility rule applies:

$$(t \times t') \textbf{ is } (u \times u') \quad \text{iff} \quad t \textbf{ is } u \quad \text{and} \quad t' \textbf{ is } u'$$

While presenting the type system of Galileo, we will discuss both the compatibility rules for the type hierarchies and the typing rules for the control structures.

2.1 Tuples

The tuple data structure, like records of programming languages and traditional database models, consists of an unordered set of <identifier (attribute or label), denotable value> pairs. A denotation of a tuple is:

```
PaulBrown :=
(Name := "Paul"
 and  Surname := "Brown"
 and  BirthDate := (Month := "Jan"
                    and  Day :=   2
                    and  Year :=  1958))
```

A tuple type consists of an unordered set of <identifier, type> pairs. For instance:

```
PaulBrown : ( Name: string
        and   Surname: string
        and   BirthDate : ( Month: string
                       and  Day: num
                       and  Year: num ))
```

The following is the general typing rule for tuple expressions:

$$E_1: t_1, \ldots, E_n: t_n$$

$$(Id_1:=E_1 \text{ and} \ldots \text{ and } Id_n:=E_n) : (Id_1: t_1 \text{ and} \ldots \text{ and } Id_n: t_n)$$

Let us consider the following concrete type definitions:

```
type Person := ( Name: string
                and Surname: string
                and BirthDate: Date)

and Student := ( Name: string
                and Surname: string
                and BirthDate: Date
                and School: string)
```

where Date is a concrete tuple type for dates. The **is** relation on tuple types is defined by the following rule:

$(\ Id_1:t_1 \text{ and} \ldots \text{ and } Id_n:t_n \text{ and} \ldots \text{ and } Id_m:t_m) \text{ is } (Id_1:u_1 \text{ and} \ldots \text{ and } Id_n:u_n)$
iff $t_i \text{ is } u_i, i=1, \ldots, n.$

That is, a tuple type t is a subtype of another tuple type u iff t has all the attributes of u, and possibly more, and the types of common attributes are respectively in the **is** relation.

According to the above rule, the type-checker can infer that Student **is** Person (and not vice versa), and thus a value of type Student can be used where a value of type Person is expected. If MarySmith is a value of type Student, it is possible to see it as a value of type Person by forcing its type: the evaluation of the expression

MarySmith : Person

yields a tuple (of type Person) in which the School attribute of students is not any longer accessible. On the contrary, if PaulBrown is a person, the expression "PaulBrown: Student" is statically detected as a type violation.

Tuples are equipped with the **of** operator which returns the value associated with an identifier:

Name **of** PaulBrown   yields   "Paul".

The typing rule for tuple selection can be stated as follows:

$$E : (Id : t)$$
$$\overline{\hspace{3cm}}$$
$$(Id \textbf{ of } E) : t$$

This rule, combined with the general rule (1) for subtyping, establishes that "Id **of** E" is well typed when E has *any subtype* of the tuple type "(Id: t)" for some type t. Thus "Name **of** P" is well typed if P is a person or a student, while "School **of** P" is well typed only if P is a student.

It is important to stress that the **is** relation on tuple types formalizes the notion of *inheritance* in the object oriented languages, like Smalltalk [GR83]. With reference to the above examples, the fact "Student **is** Person" is equivalent to: the subtype Student inherits attributes Name, Surname and BirthDate from its supertype Person. Thus, subtyping on tuples captures *attribute inheritance*. Moreover, as it will be pointed out shortly, *function inheritance* from supertypes to subtypes is the general property of type hierarchies.

The subtype rule for tuples supports also *multiple inheritance* : in the following example the type EmployedStudent happens to be a subtype of both Student and Employee (which is in turn a subtype of Person):

```
type Employee := ( Name: string
              and Surname: string
              and BirthDate: Date
              and Salary : num)

and EmployedStudent := (    Name: string
                        and Surname: string
                        and BirthDate: Date
                        and Salary : num
                        and School : string
                        and KindOfWork : string)
```

2.2 Variants

A variant type consists of a set of alternative types, discriminated by different *tags* :

```
type OfficeEmployee :=
        <Technician: (Name: string and  Skill: string)
        or Secretary: (Name: string and TypingSpeed: string)>
```

A variant value is a single pair <tag, value>, such as:

```
JohnSmith := <Secretary:=(Name:="John Smith"
                    and TypingSpeed:="High")>
```

The following typing rule holds for variants expressions:

$$\frac{E: t}{<Id:= E> : <Id: t>}$$

Thus, the denotations of variant values happen to get degenerate variant types, with just one alternative:

```
JohnSmith : <Secretary: (Name: string and TypingSpeed: string)>
```

The compatibility rule for variant types is :

$<Id_1:u_1$ **or** ... **or** $Id_n:u_n>$ **is** $<Id_1:t_1$ **or** ... **or** $Id_n:t_n$ **or** ... **or** $Id_m:t_m>$
iff t_i **is** u_i, i=1, ... , n.

That is, a variant type t is a subtype of another variant type u iff all the tags of t are also tags of u, and the types of common tags are respectively in the **is** relation.

According to the above rule, JohnSmith has also type OfficeEmployee, as well as "Day **is** WeekendDay" holds, with reference with the following example, in which variant types are used to model enumerations:

```
type Day := <Mon or Tue or Wed or Thu or Fri or Sat or Sun>
and WeekendDay := <Sat or Sun>
```

where <Sat **or** Sun> is syntactic sugar for <Sat: **null or** Sun: **null**>

The **case** construct is provided to test the tag of a variant and to bind the value associated to that tag to a local identifier:

```
case JohnSmith when
<Technician:= x. Skill of x
    or Secreatry:= y. TypingSpeed of y>          yields  "High"
```

The following is the typing rule for **case**:

$$E: <Id_1: t_1 \text{ or ... or } Id_n: t_n>, \quad x_i: t_i \Rightarrow E_i: t \text{ for each i in } 1, ... , n$$

$$(\text{case E when } <Id_1:= x_1. E_1 \text{ or ... or } Id_n:= x_n. E_n>) : t$$

Again, the combination of this rule with (1) ensures us that no error can occur in a **case** expression when working on a variant value whose type is a subtype of that of the **case** body.

2.3 Type Hierarchies and Function Applications

The type constructor "$\rightarrow$" is used to build type expressions for functions, as in "+ :
(**num** $\times$ **num**) $\rightarrow$ **num**". The following is the typing rule for function application:

$$F: t \rightarrow u , E: t$$

$$(F\ E) : u$$

Notice that F is, in general, an expression evaluating to a function value. Thus, in "+(3,4)" (i.e. "3+4" in infix notation), the type "**num** $\times$ **num**" of the argument pair is successfully matched against the type of the domain of "+". Combining this rule for function application with (1), we get that any function F with type "$t \rightarrow u$" can be safely applied to a value V with type t', provided that t' **is** t, since any value of type t' can be used as a value of type t. Let us consider the following function to concatenate Name and Surname of a person:

```
use WholeName (P: Person) := Name of P # " " # Surname of P;
```

where "#" denotes string concatenation. The type "Person $\rightarrow$ **string**" is inferred for WholeName, and thus the function can be applied to values of any subtype of Person. Consequently, type hierarchies allow that subtypes *inherit* functions from their supertypes, and support the notion of *programming by data specialization*, originally introduced in Simula 67 [DN66] and later generalized in Smalltalk and TAXIS. Complex software applications, expecially those related to databases, can be designed and implemented incrementally: once a function has been designed and tested for the most general data, it can be still used with data of any subtype introduced later on in the software development process. Moreover, new functions on the subtypes can be defined in terms of the old functions.

2.4 Sequences and Polymorphic Operators

A sequence is a finite ordered collection of homogeneus elements, i.e. values of the same type. A sequence type is denoted by type constructor **seq** followed by the type of the elements:

```
[3; 5; 2; 3]  : seq num
[<Mon>; <Thu>; <Sat>]  : seq Day
```

The **is** rule for sequences is straightforward:

$$(\text{seq t}) \textbf{ is } (\text{seq u}) \quad \text{iff} \quad \text{t is u.}$$

The following examples show some operators on sequences:

```
1 :: [2; 3; 4]          yields [1; 2; 3; 4]
first [2; 3; 4]         yields 2
rest  [2; 3; 4]         yields [3; 4]
[0; 1] append [2; 3; 4] yields [0; 1; 2; 3; 4]
3 isin [2; 3; 4]        yields true
emptyseq [2; 3; 4]      yields false

all x in [2; 3; 2; 3; 4] with x > 2      yields  [3; 3; 4]
for y in [(Name="Jim" and Age:=20); (Name="Alice" and Age:=32)]
with Age of y > 20  do Name of y         yields ["Alice"]
```

It is important to stress that the primitive operators on sequences are intrinsically *polymorphic*, i.e. they can work on sequences whose elements are of arbitrary type. The type of polymorphic primitive operators is specified using *type variables* (denoted α, β, ...) which stand for arbitrary type, e.g.:

```
first :  seq α  →α              for any type α
::    :  α × seq α  →seq  α     for any type α
```

A type which contains type variables is said *polymorphic*, while it is said *monomorphic* (or *ground*) otherwise. For instance, the type of "::" specifies that it is possible to prefix an element of arbitrary type α to a sequence whose elements are of the same type. The polymorphic operators can be applied to arguments of ground types, thus receiving different ground types in different contexts:

```
in first [1; 2; 3]      first has type   seq num →num
in first [[2; 3]; [1]]  first has type
                        seq (seq num) →seq num
```

The type system of Galileo deals with these predefined polymorphic operators through a smooth extension of the polymorphic type-checking techniques, such as those used in the functional language ML [MILN78] [CARD85b].

Polymorphic type-checking is based on an unification process over type expressions with type variables. For instance, in type-checking:

```
1 :: [2; 3]   (i.e.  :: (1, [2; 3]) )
```

the domain type of "::" (i.e. (α × **seq** α)) is matched against the type of the argument "(1, [2; 3])", which is (**num** × **seq num**). As a result, the variable α is instanced with type **num**. Repeated type unifications allow to infer types for larger expressions. It is worth noting that two or more different instantiations of a same type variable correspond to the detection of a type violation in the expression, as in:

$$1 \ :: \ [\text{"a"; "b"}]$$

where the same type variable α should be instanced both with **num** and **string**.

This restriction can be relaxed taking into account the type hierarchies. To illustrate this point, let PaulBrown and MarySmith be values of type Person and Student respectively:

```
PaulBrown := (Name := "Paul"
              and  Surname := "Brown"
              and  BirthDate := (Month := "Jan"
                                 and  Day := 2
                                 and  Year := 1958)) : Person
MarySmith := (Name := "Mary"
              and  Surname := "Smith"
              and  BirthDate := (Month := "Jun"
                                 and  Day :=  2
                                 and  Year :=  1964)
              and  School := "University of Pisa")    : Student
```

Since Student **is** Person, during the type-checking of:

$$\text{MarySmith} \ :: \ [\text{PaulBrown}]$$

the type variable α contained in the type of "::" gets two possible instance types, Person and Student, which are *compatible*, in the sense that the set of possible instances for α includes a *least upper bound* with respect to the subtype partial ordering. In the example, {Student, Person} is the set of possible instance types for α, and Person is its maximum, i.e., the most general of its possible instances. Taking the maximum as the actual instance for α, a correct typing of the expression is achieved. In the example, "MarySmith :: [PaulBrown]" receives type "**seq** Person" and evaluates to:

```
[ (Name := "Paul"
   and Surname := "Brown"
   and BirthDate := (Month:= "Jan" and Day:= 2 and Year:= 1958));

  (Name := "Mary"
   and Surname := "Smith"
   and BirthDate := (Month:= "Jun" and Day:= 2 and Year:= 1964) ]
```

Notice that the tuple MarySmith is seen as a Person in the resulting sequence. As a consequence, it is possible to build sequences with heterogeneous elements, provided that the types of the elements satisfy the above compatibility property. The system is then able to infer an appropriate type for the sequence (the most general one), thus guaranteeing a secure usage of the elements of the sequence.

The general typing rule for polymorphic operator application is expressed as follows:

$$F: t_\alpha \to u_\alpha \, , \quad E: t' \, , \quad v = \text{max of Inst}_\alpha \, (t_\alpha, t') \, , \quad t' \textbf{ is } t_\alpha \leftarrow v$$

$$F(E) \; : \; u_\alpha \leftarrow v$$

where t_α and u_α are types containing the type variable α, t' is a ground type, Inst_α (t_α, t') is the set of the possible ground instances for α when matching t_α against t', v is the most general of these instance types, and finally $t_\alpha \leftarrow v$ (resp. $u_\alpha \leftarrow v$) is the ground type obtained substituting the occurrences of α with v in t_α (resp. u_α). A type violation occurs if v does not exist. A straightforward extension of this rule is needed to cope with types where more than one type variable occurs.

It is worth noting that such a rule allows a more flexible type checking strategy for other constructs of the language, such as the conditional expression

if E **then** E$_1$ **else** E$_2$

which can be thought of, from the static semantics viewpoint, as the application of a function "**ifthenelse** : (**bool** $\times \alpha \times \alpha$) $\to \alpha$" to an appropriate argument. The above typing rule for polymorphic functions states that the types of the *then* and *else* expressions are not compelled to be equal, but simply comparable with respect to relation **is**, and the most general type is chosen as the instance for α, and thus as the type of the whole conditional. For instance:

if true **then** MarySmith **else** PaulBrown

is a well typed expression even if MarySmith and PaulBrown have different types (Student and Person resp.), and Person is the type inferred for its result. Similar considerations apply to other constructs, e.g. to **case** expressions.

Unfortunately, this type checking strategy works safely only when predefined polymorphic operators are applied to arguments with *ground* types, and this is the reason that prevents us from type-checking new user defined polymorphic functions. Indeed, the user is compelled to provide ground types for the formal parameters and the result in his own function definitions, and thus only monomorphic functions can be defined in Galileo, even if predefined polymorphic operators can be safely used in their declarations. In [CW85] a type system is presented, which fully integrates parametric polymorphism with subtyping, using bounded type quantifiers. This proposal, however, leaves unsolved the problem of type inference in such a sophisticated context, that is the type-checking process when the user does not specify completely the type of a function.

2.5 Modifiable Values

Locations are provided to introduce modifiability in the language. Locations reside in a time varying store, and are associated with values of any type, included other locations. For instance, the expression "**var** 3" denotes a new location associated with the number 3, and "**var num**" is its type. The **at** operator is provided to derefence the value associated with a location, and the assignment operator $\leftarrow$ is provided to change the value of a location in the store. Notice that the keyword **var** stands both for the operator which creates new locations and the type constructor for location types. The following is a complete list of the operators on locations.

var : $\alpha \rightarrow$ **var** α

at : **var** $\alpha \rightarrow \alpha$

$\leftarrow$: **var** $\alpha \times \alpha \rightarrow$ **null** .

Here **null** is used to model the type of assignment operations. For example:

use x := **var** 3

in (x $\leftarrow$ **at** x + 1; **at** x)   yields 4

where $(E_1; \ldots; E_n)$ evaluates all expressions E_i sequentially and returns the value of E_n.

A particular subtyping compatibility rule applies for location types:

var t **is var** u iff t = u

This stricter rule is necessary to avoid run-time type violations caused by assignment operations. For instance, let us consider the following function Foo which modifies the value of a location (of type **var** Person) assigning it the Person tuple associated to PaulBrown:

    use Foo (P : **var** Person) : Person := P $\leftarrow$ PaulBrown

If, according to a coarser compatibility rule, the expression

    use Loc := **var** MarySmith **in** Foo(Loc)

would be considered well typed (recall that MarySmith is a tuple of type Student, and hence Loc: **var** Student), then the location Loc will change dynamically its type becoming a Person location, and so the School attribute of Student tuples has been lost! Thus, any subsequent use of Loc as a Student location, as in "School **of at** Loc" will generate a run-time type error.

2.6 Functions

In Galileo, functions are ordinary denotable values, which can be passed as arguments to (or returned as result from) other functions and as well be used as components of data structures. As already mentioned, functional types are built using the type constructor $\rightarrow$. Function values are denoted with the construction "**fun** (x : t) : u **is** E", with typing rule:

$$x : t \Rightarrow E : u$$
$$\overline{\textbf{(fun} (x : t) : u \textbf{ is } E) : t \rightarrow u}$$

The construct "F (x : t) : u := E" (equivalent to "F := **fun** (x : t) : u **is** E") is preferred in declaring a function F with formal parameter x and body E. We have already discussed the typing rules for function application. Here it is interesting to see how type hierarchies extend to functional types. Suppose BestStudent is a function that returns the top student of the University of Pisa in a given year (here years are represented as numbers):

$$\text{BestStudent} : \mathbf{num} \rightarrow \text{Student}$$

We can say that BestStudent returns also persons, since all students are persons. In general, all Student-valued functions are also Person-valued functions, so that we can say:

$$t \rightarrow \text{Student } \mathbf{is} \ t \rightarrow \text{Person}, \quad \text{for any type t}$$

since Student **is** Person. Now consider a function Age which returns the age of a Person tuple:

$$\text{Age} : \text{Person} \rightarrow \mathbf{num}$$

Since all students are persons, we can use this function to compute the age of a student. In general, any function on persons is also a function on students, so that we can say:

$$\text{Person} \rightarrow t' \ \mathbf{is} \ \text{Student} \rightarrow t', \quad \text{for any type t'.}$$

Combining these statements together, one gets that, given a function F with type "u $\rightarrow$ u'", a subfunction of F is one obtained either strengthening the range u' of F or broadening the domain u of F. Thus, the general compatibility rule for functional types is the following:

$$t \rightarrow t' \ \mathbf{is} \ u \rightarrow u' \quad \text{iff} \quad u \ \mathbf{is} \ t \ \text{ and } \ t' \ \mathbf{is} \ u'.$$

For instance, a "Person $\rightarrow$ Student" function can be used where a "Student $\rightarrow$ Person" function is expected. This rule is all we need for the type checking of higher order functions, i.e., those which have functions as arguments or which returns functions as result.

2.7 The types "any" and "none"

It is possible to see the subtype partial ordering over types as a complete lattice, since a most general type **any** and a most particular type **none** exist. The following compatibility rules hold:

$$t \ \mathbf{is \ any}, \quad \text{for any type t}$$
$$\mathbf{none \ is} \ t, \quad \text{for any type t.}$$

any is the set union of all types, i.e., the type of all values, equipped with no operator, since it inherits functions from no type. On the other hand, **none** is the set intersection of all types, equipped with all the operators of the language, since it inherits functions from any type. It is worth noting that **none** is not empty: the undefined value $\bot$, which, in the semantics, is the value denoted by a diverging or failing expression, is a value of any type, since expressions of any type can diverge or fail. Thus, **none** is the singleton set whose only element is $\bot$, and all operators can be applied to $\bot$ yielding $\bot$ as a result, since Galileo has a call-by-value (strict) semantics.

Thus, it is natural to see **none** as the type of explicit failure expressions "**failwith** String", which are used to generate run time failures when exceptional situations occur during the execution. The string associated to a failure expression is used in the failure handling mechanism of the language.

$$\frac{E : \textbf{string}}{(\textbf{failwith } E) : \textbf{none}}$$

The following function checks whether a number is even, and generates a failure if the number is not an integer.

```
Even (n : num) : bool :=
                      if Integer(x) then n mod 2 = 0
                      else failwith "Not an integer"
```

Here, the *then* expression gets type **bool** and the *else* failure expression gets type **none**, so that the whole function body gets type **bool**, i.e., the most general type among **bool** and **none**.

The **any** type can be used to mimic a class of simple polymorphic functions, such as the following function Length which counts the elements of a sequence of arbitrary type:

```
rec Length (L : seq any) : num :=
           if emptyseq L then  0
           else 1+ Length(rest L)
```

Another possible use of the **any** type is for protection purposes, e.g. in hiding some components of a data structure by forcing it a type with appropriate **any** components. In this example **any** is used to hide the BirthDate attribute of the person PaulBrown:

```
PaulBrown : (Name:string and Surname:string and BirthDate: any)

yields    (Name:="Paul" and Surname:="Brown" and BirthDate:= - )
```

Notice that the system represents with a hyphen the **any** values, in order to stress that they can be anything, and that no primitive operator is available on them.

3 ABSTRACT TYPES

The type of the values presented so far depends only on the structure of the values, i.e., a *structural equivalence* rule is adopted. In contrast, a user-defined abstract type is always different from all other concrete or abstract types, as well it differs from its representation type, i.e., a *name equivalence* rule is adopted for abstract types. New abstract types are introduced with the following declaration:

$$\textbf{type } Id \Leftrightarrow RepType \quad \{\textbf{assert } BoolExp\}$$

This declaration introduces the following bindings: **a)** Id is bound to a new abstract type whose domain is isomorphic to the domain of the representation type RepType, possibly restricted by the assertion BoolExp; **b)** the identifiers mkId and repId are bound to two primitive functions, automatically declared, to map values of the representation type into the abstract type and vice versa:

$$\text{mkId} : \text{RepType} \rightarrow \text{Id}$$
$$\text{repId} : \text{Id} \rightarrow \text{RepType}$$

If an **assert** clause is present, BoolExp is a boolean expression on the values of RepType, which imposes constraints on the values of the abstract type. These constraints are controlled at execution time, when a value is created by a mkId operation. The following is the declaration of an abstract tuple type for dates. The keyword **this** in the assertions is bound to the argument of mkId. Some self explanatory abbreviations are used to simplify the assertions.

```
type Date ⇔ (Month: num
               and Day: num
               and Year: num )
    assert
      use this in
        Month within (1, 12) And
        Day within (1, 31) And
        Year within (1900, 2000) And
        Day ≤ (if Month=2
            then if Year mod 4 = 0 then 28 else 29
            else if Month isin [4; 6; 9; 11] then 30 else 31)
```

Thus, "mkDate(Month:=2 **and** Day:=15 **and** Year:=1985)" denotes a value of the new type Date, while "mkDate(Month:=2 **and** Day:=31 **and** Year:=1985)" will generate a run time failure because of the assertion on dates. To define an abstract type with hidden representation, but with user defined operations, the **with** construct might be used, as in:

```
type Time ⇔ (Hrs: num
               and Mins: num)
      assert use this in Hrs within (0,23) And Mins within (0,59)
with    Hours (t: Time) : num := Hrs of repTime (t)
and     Minutes (t: Time) : num := Mins of repTime (t)
and     MakeTime (x: num, y: num) : Time :=
                      mkTime (Hrs:= x  and  Mins:= y)
```

This declaration exports an abstract type Time together with three functions Hours, Minutes and MakeTime. The two primitive operations mkTime and repTime are available only in the **with** part, but they are not exported in the scope of the declaration. In this way, it is possible to tailor particular operations for each type, which cannot be used for values of other types. Once defined, new abstract types have the same status as primitive types like **num** or **bool**: indeed, primitive types can be consistently regarded as predefined abstract types provided by the language.

3.1 Abstract Types and Type Hierarchies

To enforce protection on abstract types, the subtype relation between abstract types must be explicitly declared to the type checker when defining a new type:

$$\textbf{type Id is } T_1, \ldots, T_k \Leftrightarrow \text{RepType}$$

where each T_i is either another abstract type with representation type $RepType_i$, and in this case "RepType **is** $RepType_i$" must hold, or a primitive type, like **num**, and in this case "RepType **is** T_i" must hold. With this declaration the new type Id is considered as a subtype of each T_i, and it inherits the assertions of each T_i.

```
type Person  ⇔
              (Name:  string
               and Surname:  string
               and BirthDate:  Date)

and Student  is Person  ⇔
              (Name:  string
               and Surname:  string
               and BirthDate:  Date
               and School:  string)
```

The following abbreviation makes evident that the subtype Student inherits attributes from the type Person:

```
type Person  ⇔
              (Name:  string
               and Surname:  string
               and BirthDate:  Date)

and  Student  ⇔
              (is Person
               and School:  string)
```

3.2 Operator Overloading

A further form of inheritance is provided through a declarative form of operator overloading, available when defining abstract types with **import** clauses:

$$\textbf{type } Id \Leftrightarrow RepType \textbf{ import } Op_1, \ldots, Op_n$$

where each Op_i is an identifier bound to an operator on values of the representation type. The system overloads these identifiers, also binding them with new operators which work with values of the new type Id. For instance, consider the following abstract type for weights:

$$\textbf{type } Weight \Leftrightarrow \textbf{num} \ \ \textbf{assert this} \geq 0 \ \ \textbf{import } +, -$$

As a consequence of this declaration, the new type Weight inherits the operators specified in the **import** clause, in the sense that the identifiers "−"and "+" are bound to two different functions: e.g., "−" is bound both to the primitive number subtraction (− : $\textbf{num} \times \textbf{num} \to \textbf{num}$) and to the overloaded weight subtraction (− : Weight $\times$ Weight $\to$ Weight) that has been automatically declared by the system as follows:

```
−   (x: Weight, y: Weight) : Weight :=
          mkWeight(repWeight(x) - repWeight(y))
```

Notice that the weight subtraction is defined in term of the original number subtraction, but moreover it controls the assertion on weights, thus guaranteeing that weight subtraction yields proper weight values. Therefore, the expression:

```
use x:=mkWeight 4
and y:=mkWeight 5 in  x-y
```

yields a run-time failure, since weights are to be non negative. In the application of overloaded identifiers, the type checker is able to choose the right operator according to the type of the arguments: in the example above, the weight subtraction is chosen, since the identifier "−" is applied to a weight pair. There is a simple requirement on the operators which can appear in an **import** clause, namely that they should have at least an argument of the representation type. The overloading mechanism is particulary useful when an imported operator yields a value of the representation type as its result, since in this case the overloaded operator will yield values of the new abstract type with implicit assertion control, as in the example above.

This notion of overloading is quite different from subtyping inheritance. Suppose we had declared the type Weight in the following way:

$$\textbf{type } \texttt{Weight is num} \quad \Leftrightarrow \quad \texttt{num} \quad \textbf{assert this} \geq 0$$

then, although the subtraction operator can work also on weights, the result of a weight subtraction is always a number, and thus no assertion control is performed on it. However, subtyping and overloading can be freely combined when defining abstract types, as in the following definition of the type **int** of integers, which is indeed a predefined abstract type of the language:

```
type int is num  ⇔  num
        assert Integer(this)
        import +, -, /, *, =, ^
```

where Integer is a predicate yielding true on numbers with no decimal part. As a consequence of this definition, integers inherit the overloaded operators on numbers, as well as an integer can appear in the context where a number is expected. Consequently, the expression "x/y" can receive the following typings, according to the type of x and y:

```
x: num, y: num   =>  /: num×num→num  and  (x/y): num
x: num, y: int   =>  /: num×num→num  and  (x/y): num
x: int, y: num   =>  /: num×num→num  and  (x/y): num
x: int, y: int   =>  /: int×int→int  and  (x/y): int
```

Notice that in the last case, "x/y" is well typed both with the number division and with the overloaded division (which checks that the result is again an integer), because of subtyping. The system, in such cases, privileges overloaded operators, since they provide a richer information on the result. Precisely, when more than one operator is consistently applicable, the system chooses the one with most specialized domain. With reference to the above example, type **int** is then equipped with its own operators, such as proper integer division and modulus (div and mod), rounding from numbers to integers, and so on. As a result, no fuzzy coercion rules are to be stated to understand when an integer is used as an actual integer or as a real number, the type checker being able to infer proper typing under a quite general rule. Of course, explicit round or truncate operations are to be applied to use numbers in integer contexts, since a number is not, in general, an integer. As an abbreviation, the syntax "3" can be used instead of "mkint 3", and so both "3: **int**" and "3: **num**" hold, while only "3.14 : **num**" holds.

Finally, a special declaration is provided for defining abstract types which import all the primitive operators of the representation type:

$$\textbf{type } \text{Id} \leftrightarrow \text{RepType}$$

where RepType is expected to be a primitive or concrete type, in order to identify the set of the operators which are to be imported. Thus, a more concise declaration for integers is the following, which overloads all the operators on numbers:

```
type int is num ↔ num    assert   Integer(this)
```

As another example, the following is a redefinition of the abstract tuple types Person and Student:

```
type Person ↔
       (Name: string
       and Surname: string
       and BirthDate: Date)

and  Student ↔
       (is Person
       and School: string)
```

In the scope of this declaration, the expression "Name **of** JohnSmith" makes sense even if JohnSmith is an abstract value of type Person or Student, since these types are declared with the "$\leftrightarrow$" abstract type constructor, thus importing the **of** operator on tuples.

As a last comment, notice that the declarations via "$\leftrightarrow$" are nothing but a programming facility designed to avoid the syntactic burden of explicit mkId and repId operations when working with abstract types which require *no representation hiding*. For instance, weights and heights can be both represented as numbers and it is probably useful to work on weight and height abstract values in the same way one works on numbers, while the system guarantees that weights and heights are not inconsistently mixed in expressions. Moreover, abstract types without representation hiding are often useful in conceptual modeling, as we will shortly show.

4 SUPPORTING THE SDM ABSTRACTION MECHANISMS

Classes form the mechanism to represent a database by means of sequences of modifiable interrelated values. An element of a class is a value which is a unique computer representation of some entity in the world that is being modeled.

A class is characterized by a *name*, which denotes the elements of the class currently present in the database, and a *type*, which gives the structure of its elements. The type of a class must be *abstract*, since two elements of different classes are always of different types, although they may be defined with the same representation. In most cases, the structure of the elements of a class is given by means of an abstract tuple type, since aggregation is used for modeling both properties of elements and associations which relate elements. Thus, although the type of a class must be abstract, its representation must not be hidden, since it is part of the knowledge represented in a conceptual schema.

Hence, the abstract type definitions via "$\leftrightarrow$" with operator overloading are quite appropriate to describe the structure of elements of classes.

```
Persons class
    Person ↔
        (Name: string
        and Surname: string
        and BirthDate: Date)
```

This definition introduces the class identifier Persons bound to a modifiable sequence of Person values. The class extension varies in time: when a person is created with the mkPerson operator, it belongs automatically to the class, as long as it is not explicitly removed. The overloading mechanism allows the use of the **of** operator on the elements of the class Persons.

While classes and abstract types with overloading are provided to support, respectively, the extensional and intensional aspects involved in classification, *subclasses* and *type hierarchies* are provided to support, respectively, the extensional and intensional aspects involved in generalization. A subclass contains, in general, a subset of the elements of each its superclass, while the type of the elements of a subclass must be a subtype of the type of each superclass, thus modeling attribute inheritance.

```
Students subset of Persons class
    Student ↔
        (is Person
        and School: string)
```

With this declaration, the class Students contains those persons which have been included explicitly in it with an appropriate operator. Other ways for defining subclasses are by partitioning a class, or by restricting the extension of a class to those elements which satisfy some predicate.

It is important to stress that the distinction of the extensional and intensional notions involved in the SDM abstraction mechanisms increases the modeling capabilities of the language, since it allows the use of type hierarchies or abstract types with overloading independently from classes and subclasses.

5 CONCLUSIONS

The type system of Galileo has been presented, which has been designed to support the intensional notions underlying the SDM abstraction mechanisms. The type system has been implemented and it is fully operational [AOO86]. Relevant novel features are:

a) the notion of type hierarchies, which are extended to all the types of the language, and

b) an abstract data type definition capability, with assertions to restrict the values of the representation type, which allows the inheritance of operators from the representation type by means of a declarative form of overloading.

We believe that these mechanisms have proven to be useful in general and not only in conceptual modeling applications, since they allow a smooth type-checking of aspects of programming languages which are traditionally hard to handle in a sound type system, such as failure handling and coercion rules between primitive types. Further work is required to understand how conceptual modeling, and programming in general, will be affected by the full opportunities offered by a type system based on type hierarchies.

Moreover, it is a challenge to devise an inferential type system which fully integrates type hierarchies and overloading with parametric polymorphism.

Another relevant feature of the language, which has not been discussed here and which it is not yet available in the present implementation, is values persistence. Galileo has been designed as a persistence conceptual language, i.e., a language with values persistence as an orthogonal property of the type system [AGO85]: instead of having persistence as a property of values of special types, such as files of programming languages, it is assumed that persistence is a property of any value that is accessible from the top level environment.

REFERENCES

[ACO85] [AGO85]: [ALBA83] [AOO86] [BMS84] [BMW80] [CARD84] [CARD85b] [CW85] [DN66] [GR83] [MILN78]

Chapter 9

Integrating Data Type Inheritance into Logic Programming

Hassan Aït-Kaci[1]
Roger Nasr
MCC, AI Program

ABSTRACT *This document describes a particular facet of an experimental programming model whose design is the objective of our on-going research at MCC - the $\mathcal{E}$ programming environment. $\mathcal{E}$ draws from the most recent theory and technology in typed relational and functional programming. We start by making general observations meant to put our work in context within the research community, and proceed to summarize that particular sub-language of $\mathcal{E}$ which combines type inheritance and logic programming - a prototype named LOGIN.*

Keywords: Object Inheritance, Object-Oriented Programming, Logic Programming, Functional Programming, Data Types, Information Storage and Retrieval.

> Pourquoi faire simple quand on peut faire compliqué?
> LES SHADOKS - *French TV Cartoon Characters.*

1. OVERVIEW

The $\mathcal{E}$ project is a component of the ISA Project in the AI Program. It pursues research in architecture design for advanced programming environments. It aims to develop a virtual machine design whose model of computation, data representation, and process communication are based on principles evolved out of the latest technology and research in programming language and computation theory. Our ground assumption is that it is now economically feasible to envision integration and distribution of very sophisticated processors and storage devices. As a result, it has become feasible to subordinate hardware design to that of very high-level software, rather than the converse as was conventionally thought until recently.

---

1 This research is being carried out as part of the $\mathcal{E}$ programming language project by the Intelligent Systems Architecture [AN486].

Understanding that existing languages or data models are better suited for particular tasks, we believe that the sensible thing to do is to attempt pulling together from the better ones their "nice" features while eliminating as much as possible their "bad" features. Even further, research on ε tries to automate the task of deciding which particular control structure manages a given execution better, or how a particular data structure must be implemented. The gain is a enhancement of programming correctness confidence, as well as systematic optimization potential.

We are concentrating our efforts at a level between specific physical architecture topologies, and high-level AI applications. The idea is to develop an advanced abstract machine (instruction set) where high-level operations commonly used in such applications are cleanly integrated as high-performance primitives. This entails developing a compiler (translator and code improver) which will carry out the mapping from high-level programming language constructs into the appropriate machine instructions. This is why ε is conceived as a logic [ALN87]. In particular, the very ε system is being designed in ε, starting from a well-defined kernel. The system is thus better construed as a meta-compiler.

Three computation paradigms have evolved out of AI: *functional programming*, *logic programming*, and *type inheritance*. The first one, in the form of Lisp, emerged as the deed of an AI pioneer - the very same who coined the phrase *"Artificial Intelligence."* The second one, in the form of Micro-Planner and later Prolog, came directly from Automatic Theorem-Proving research. Finally, frames and semantic network technology [MINS75] [FIND79] is the latest AI invention which has reluctantly been acknowledged as a useful tool in the broader area of Computer Science.

In all three cases, a sound and well-understood formal ground was to justify the behavior of the corresponding languages, and thus allow the spectacular optimization of their implementation. Loosely designed after Church's formal combinatory logic [CHUR40], Lisp is now rendered as well streamlined engines; so is Prolog, based on SLD-resolution, thanks to D.H.D Warren's ingenuity [WARR83]; and Inheritance, based on a logic of types [AITK84a], is showing its power as a very efficiently compilable computation model [ABN85], not to mention its intriguing promise as a massively parallel hardware model.

Our idea, as always, is simple - ε is to be the composition of three sublanguages corresponding to pairwise combinations of each of the three chosen models:

LOGIN
Logic and Inheritance - is a typed Prolog where the power of inheritance among partially-ordered type objects is exploited through type unification [AN86b].

Le FUN
Logic, equations, and Functions - is a typeless extension of Prolog which can deal with user-definable functions, and this, without explicit annotations [AN87b].

FooL
Functional Object-Ordered Language - is a typed functional language where types are partially-ordered objects and "first-class citizens" [AN87a].

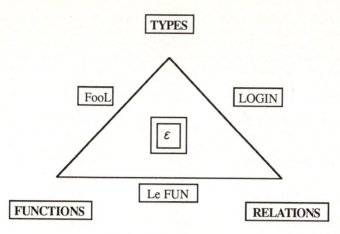

Figure 1: The ε Triangle

Figure 1 illustrates the above approach by a triangle whose vertices are the given computation models, and the edges the sought pairwise combinations. Moving toward the barycenter of this triangle will produce ε .

The present contribution is a summary of LOGIN [AN87a]. The rest of ε (*vz.*, FooL, Le FUN, and LIFE) is described somewhere else [ALN86] [AN86c] [AN87a]. So are applications [ANS86].

We introduce in Section 2 the flavor of what we believe to be a more expressive and efficient way of using taxonomic information, as opposed to straight Prolog. Then, in Section 3, we discuss the adequacy of the use of first-order terms in Prolog. This leads to a proposal in Section 4 to embody taxonomic information as record-like type structures. Finally, in Sections 5 and 6, a simple interpreter is illustrated on examples, showing how this extended notion of terms may be integrated into the SLD-resolution mechanism of Prolog.

2. MOTIVATION

Inheritance can be well captured in logic by the semantics of logical implication. Indeed,

$$\forall\, x.\ Whale(x) \Rightarrow Mammal(x)$$

is *semantically* satisfactory. However, it is not *pragmatically* satisfactory. In a first-order logic deduction system using this implication, inheritance from "mammal" to "whale" is achieved by an *inference* step; whereas, the special kind of information expressed in this formula somehow does not seem to be meant as a deduction step - thus *lengthening* proofs - but rather, as a means to accelerate, or focus, a deduction process - thus *shortening* proofs.

Many proposals have been offered to deal with inheritance and taxonomic information in Automated Deduction. Admittedly, expressing everything in first-order logic as proposed in [DK79], and [AF82], is semantically correct. Nevertheless, these approaches dodge the issue of improving the operational deduction process. Other more operational attempts, like those reported in [MM79], and [BFL83], propose the use of some forms of semantic network. However, it is not always clear what semantics to attribute to these formalisms which, in any case, lose the simple elegance of Prolog-like expressiveness.

As shown in [AITK84a], the syntax and operational interpretation of first-order terms can be extended to accommodate for taxonomic ordering relations between constructor symbols. As a result, we propose a simple and efficient paradigm of unification which allows the separation of (multiple) inheritance from the logical inference machinery of Prolog.

Let us consider the following example:

It is known that all persons like themselves. Also, students are persons, and the individual John is a student.

This simple information can be expressed in first-order logic as:

$$\forall x.\, person(x) \Rightarrow likes(x,x)$$
$$\&\quad \forall x.\, student(x) \Rightarrow person(x)$$
$$\&\quad student(john).$$

and thus in Prolog by:

```
likes(X,X)  :- person(X).
person(X)  :- student(X).
student(john).
```

To check whether John likes himself is hence:

```
?- likes(john,john).
Yes
```

On the other hand, we can equivalently represent the information above in *typed* first-order logic as follows:

$$\forall x. \in person.\, likes(x,x)$$
$$\&\quad student \subset person$$
$$\&\quad john \in student.$$

Now, if *type checking* (*i.e.*, that one set is the subset of another, or one element belongs to a set), can be done efficiently, then the typed translation can achieve better performance, with no loss of semantics. Indeed, in our little example, to infer that John likes himself is immediate - one application of *modus ponens* rather than two in this case. This simple idea can be made practical, and is the basis of the extension of Prolog we are presenting in this document.

In Prolog, a resolution step involves a state transition with context saving, variable binding, *etc.*, and is therefore costly. The simple kind of logical implication in the above problem should not contribute to the length of a deduction. Indeed, in this example it should be immediate to see that by virtue of being a student John is also a person. It would be convenient if one could *declare* that, in the unification process, the symbol *student* can match the symbol *person*. Such declarations could look like

```
student < person.
john < student.
```

where the symbol '<' stands for "is a".

In view of these declarations, the original problem could be reformulated using a typing notation such as

```
likes(X : person, X).
```

Then, by a *unification* step rather than by a *resolution* step, the previous answer follows for the query:

```
?- likes(john,john).
```

Thus, we can view the process of unification as computing a greatest lower bound of two symbols relative to the < ordering.

This simple example may not be convincing as a true improvement. That is, one may argue that trading a unification step for a resolution step is not worthwhile. However, unification with inheritance is by far more economical than resolution. Even if this were not the case - *i.e.*, the two steps of computations were equally costly - as the length of inheritance chains increases the motivation for using fast unification rather than resolution appears more clearly.

3. AN OPERATIONAL INTERPRETATION OF TERMS

In the logic interpretation of Prolog, functional first-order terms which are not variables appear as Skolem constants or functions. However, in Prolog, such *functions* are never evaluated. Rather, they are used *operationally* as *type constructors*. The most known example is the famous *cons* list constructor; but a Prolog user can take advantage of this operational interpretation of terms for organizing data, as for instance, in database applications of Prolog [KOWA78].

As a result, Prolog's operational use of first-order terms makes them behave as *record structures; e.g.*, a term as *person(x,y,z)* is seen as a three-field record, whose fields may be given some conventional interpretation by the programmer (say, first argument is name, second is date of birth, and third is sex). The implicit operational semantics of such a constructor term is that it denotes the set (type) of all "person" records in the database.

Thus, unification of first-order terms becomes a simple-minded *inheritance* operation, as variables in terms act as slots which are filled as they become instantiated. A subtype of

$$person(x,y,z)$$

a generic denotation of the set of all (records of) persons in the database, may thus be

$$person(name(john,x),y,male)$$

a generic denotation of the set of all (records of) male persons with first name John in the database. In fact, under this interpretation of terms as types, unification is interpreted as *intersection* of types. For example, the intersection of the set of persons whose last name is the same as their first name

$$person(name(x,x),y,z)$$

with the set of all male persons whose first name is John

$$person(name(john,x),y,male)$$

must be indeed the set of all male persons named John John

$$person(name(john,john),x,male).$$

Since they are not operationally used as functions, Prolog first-order terms suffer from undeserved limitations in their syntax, a legacy of their original functional semantics.

Looking at first-order terms purely syntactically in their use as type constructors, one finds that *fixed arity* of signature symbols is an irrelevant burden. For example, if after extensive use of a three-field record $person(x,y,z)$, a user realizes that a fourth field (say, social security number) is needed, all previous occurrences of the *person* record must be revised and given a fourth argument [DK79].

Another limitation, which is also a corollary of fixed-arity, is that the interpretation of argument positions is non-transparent to the user. Indeed, in using a *person* record, one must always be aware that the first argument is a name, the second is a date, *etc*. Clearly, the classical explicit labeling of record fields by symbolic keywords is better than implicitly limiting these labels to be ordered ungapped sequences of integers.

The third most fundamental limitation of terms as type structures can be best understood when one ponders upon the respective roles of signature and variable symbols in term unification. A signature symbol is a type constructor and thus acts as an instantiation *filter*. Indeed, unification fails for two non-identical signature symbols. As a result, any further instance of, say, $person(x,y,z)$ must have *person* as root symbol. There is no reason why this filtering role of constructor symbols must be limited to an open/closed behavior. Indeed, $person(x,y,z)$ should be allowed to be further instantiated as $student(x,y,z)$ if the interpretation of the data is such that a "student" type is a subtype of a "person" type. This gradual filtering can be expressed as a partial ordering (type subsumption) on the constructor signature. Hence, unification of signature symbols is now seen as a *greatest lower bound* (GLB) operation. If a signature is augmented with a special *least element* symbol ⊥ denoting failure of unification, conventional unification of constructor symbols is still a GLB operation.

On the other hand, a variable occurrence means the absence of filter; *i.e.*, it is a *wild card* for term instantiation. As importantly, a variable has a second role as it acts as a tag imposing *equality constraints* among subterms - *all* occurrences of the same variable in a given term must be instantiated by identical terms. As an instantiation wild card, a variable behaves as a filter - very permissive, but a filter nonetheless. As a tag, it behaves as an equality constraint. It is a key observation that variables should not carry such a dual information. Firstly, because it is the role of the signature symbols to carry filtering information. And secondly, because even in their equational role, variables are unduly limited. Indeed, as variables are allowed to occur only as leaves in a term, they cannot impose equality constraints *within* the term; *i.e.*, anywhere from root to leaves.

Based on these observations, it is natural that the wild card role should be played by a special *greatest element* symbol T augmenting the signature. As for equality constraints, we propose that variables be called *tags* and allowed to appear anywhere within a term.

All the foregoing limitations are overcome in the syntax of partially-ordered type structures defined next.

4. A CALCULUS OF TYPE CONTAINMENT

We shall call the syntactic representation of a structured type a *ψ-term*. Informally, a *ψ*-term consists of:

1. A *root symbol* which is a type constructor, and denotes a class of objects.

2. *Attribute labels* which are record field symbols, associated with *ψ*-terms. Each label denotes a function *in intenso* from the root type to the type denoted by its associated sub-*ψ*-term. Concatenation of labels denotes function composition.

3. *Coreference* constraints among paths of labels which indicate that the corresponding attribute compositions denote the same functions. In other words, coreference specifies that some functional diagram of attributes must be commutative.

Consider an example of such a *ψ*-term where the root symbol is *person*:

```
person(id => name;
       born => date(day => integer;
                     month => monthname;
                     year => integer);
       father => person)
```

It has three sub-*ψ*-terms under the attribute labels *id*, *born*, and *father*, respectively. We follow the convention of using identifiers starting with a lower-case letter for type symbols and attribute labels. Identifiers starting with an upper-case letter are *tag* symbols and denote coreference among attribute compositions. An example of a *ψ*-term with tags is:

```
person(id => name(first => string;
                  last => X : string);
       father => person(id => name(last => X : string)))
```

The tag symbol *X* occurs under *id.last* and *father.id.last*, and indicates a coreference constraint; *i.e.*, identical substructures.

To be consistent, a *ψ*-term's syntax cannot be such that different type structures are tagged by the same tag symbol. For example, if something other than *string* appeared at the address *father.id.last* in the above example, the *ψ*-term would be ill-formed. Hence, in a well-formed *ψ*-term - or *wft*, for short - we shall omit writing more than once the type for any given tag. For instance, the above *ψ*-term will rather be written as:

```
person(id => name(first => string;
                  last => X : string);
       father => person(id => name(last => X))).
```

In particular, this convention allows the concise representation of *infinite* structures such as:

```
person(id => name(first => string;
                  last => string);
       father => X : person(son => person(father => X)))
```

where a cyclic coreference is tagged by *X*.

The type signature Σ is a partially-ordered set of symbols. Such a signature always contains two special elements: a greatest element (T) and a least element ($\perp$). Type symbols denote sets of objects, and the partial order on Σ denotes set inclusion. Hence, T denotes the set of all possible objects - the *universe*. We shall omit writing the symbol T explicitly in a wft; by convention, whenever a type symbol is missing, it is understood to be T. For example, in the wft:

```
person(id => (first => string;
             last => X);
       father => person(id => name(last => X))).
```

T is the type symbol occurring at addresses *id*, *id.last*, and *father.id.last*.

On the other hand, $\perp$ denotes the empty set and is the type of no object. Consequently, $\perp$ may appear in no wft other than $\perp$, since that would entail that there be no possible object for the corresponding attribute. As a result, any ψ-term with at least one occurrence of $\perp$ is identified with $\perp$. Finally, since the information content of tags is simply to impose coreference constraints, it is clear that any one-to-one renaming of a wft's tags does not alter the information content of the wft. In summary, we shall consider wfts satisfying the above syntactic limitations to be structured type expressions.

The partial order on the type signature Σ is extended to the set of wfts in such a way as to reflect set inclusion interpretation. Informally, a wft t_1 is a *subtype* of a wft t_2 if:

1. The root symbol of t_1 is a subtype in Σ of the root symbol of t_2;

2. all attribute labels of t_2 are also attribute labels of t_1, and their wfts in t_1 are subtypes of their corresponding wfts in t_2; and,

3. All coreference constraints binding in t_2 must also be binding in t_1.

For example, if Σ is such that *student* < *person* and *austin* < *cityname*, then the wft

```
student(id => name(first => string;
                   last => X : string);
        lives_at => Y : address(city => austin);
        father => person(id => name(last => X);
                         lives_at => Y))
```

is a subtype of the wft

```
person(id => name(last => X : string);
       lives_at => address(city => cityname);
       father => person(id => name(last => X))).
```

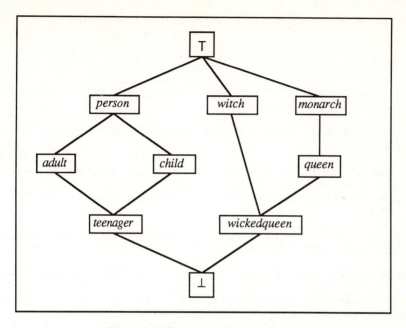

Figure 2: A Signature which is a lattice

In fact, a stronger result is proved in [AITK84a]. Namely, if the signature Σ is such that GLBs (respectively, LUBs) exist for all pairs of type symbols with respect to the signature ordering, then GLBs (LUBs) also exist for the extended wft ordering. In other words, the wft ordering extends a (semi-)lattice structure from the signature to the wfts. As an example, if we consider the signature of Figure 2, then the LUB of the wft

```
child(knows => X : person(knows => queen;
                                hates => Y : monarch);
        hates => child(knows => Y;
                        likes => wicked_queen);
        likes => X)
```

and the wft

```
adult(knows => adult(knows => witch);
        hates => person(knows => X : monarch;
                        likes => X))
```

is the wft

```
person(knows => person;
        hates => person(knows => monarch;
                        likes => monarch))
```

and their GLB is the wft

```
teenager(knows => X : adult(knows => wicked_queen;
                                hates => Y : wicked_queen);
           hates => child(knows => Y;
                          likes => Y);
           likes => X).
```

The conventional case of first-order terms is just a particular restriction of the above. Namely, first-order terms are ψ-terms such that:

1. the signature is a *flat* lattice - *i.e.*, such that all the symbols, except for T and ⊥, are incomparable;

2. tags may appear only at leaf level, and when so, only with the symbol T; and,

3. attribute labels are fixed initial sequences of natural numbers for each signature symbol.

Furthermore, the GLB is given by first-order unification [ROBI65], and the LUB is given by first-order generalization [REYN70].

For the purpose of integrating logic programming and inheritance, all we shall need is ψ-term unification. We shall assume that the signature is a lower semi-lattice - *i.e.*, GLBs exist for all pairs of type symbols. We need an algorithm which, given any pair of wfts, will compute the greatest wft which is a subtype of both wfts.

A ψ-term unification algorithm was originally given and proven correct in [AITK84a], and further detailed in [AN86b] (see also [AN86a]). It uses the same idea as the method used by Huet [HUET76] for unification of regular first-order terms based on a fast procedure for congruence closure. However, Huet's algorithm is devised for conventional (*i*) fixed-arity terms with (*ii*) arguments identified by position, and (*iii*) over flat signatures. The ψ-term unification algorithm does not impose these stringent restrictions.

5. A SIMPLE EXAMPLE

LOGIN is simply Prolog where first-order terms are replaced by ψ-terms. Thus, we shall simply show that the skeleton of a Prolog interpreter implementing a top-down/left-right backtracking search procedure can be adapted in a straightforfard manner. The unification procedure is simply replaced by ψ-term unification, altered to allow for undoing coreference merging and type coercion upon backtracking.

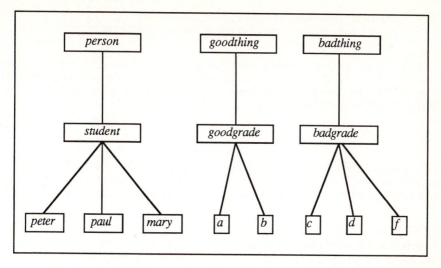

Figure 3: A Signature for the Simple Example

Let us consider the following simple example. The signature of Figure 3 is declared in LOGIN by

```
student < person.
{peter, paul, mary} < student.
{goodgrade ,badgrade} < grade.
goodgrade < goodthing.
{a, b} < goodgrade.
{c, d, f} < badgrade.
```

This essentially expresses the facts that a student is a person. Peter, Paul, and Mary are students. Good grades and bad grades are grades. A good grade is also a good thing. 'A' and 'B' are good grades; but, 'C', 'D', and 'F' are bad grades.

In this context, we can define the following facts and rules.

It is known that all persons like themselves. Also, Peter likes Mary; and, all persons like all good things. Thus, in LOGIN,

```
likes(X : person, X).
likes(peter,mary).
likes(person,goodthing).
```

Peter got a 'C'; Paul got an 'F', and Mary an 'A'. Thus,

```
got(peter,c).
got(paul,f).
got(mary,a).
```

Lastly, it is known that a person is happy if she got something which she likes. Alternatively, a person is happy if he likes something which got a good thing. Hence,

```
happy(X : person) :- likes(X,Y),got(X,Y).
happy(X : person) :- likes(X,Y),got(Y,goodthing).
```

Mary is happy because she likes good things, and she got an 'A' - which is a good thing. She is also happy because she likes herself, and she got a good thing. Peter is happy because he likes Mary, who got a good thing. Thus, a query asking for some "happy" object in the database will yield:

```
?- happy(X).
X = mary
```

The way LOGIN finds the first answer ($X = mary$) is as follows. The invoking resolvent *happy*(X : T) unifies with the head of the first defining rule of the *happy* procedure, by coercing X : T to X : *person*. The new resolvent thus is:

$$likes(X : person, Y), got(X, Y). \qquad (1)$$

Next, *likes*(X : *person*,Y) unifies with the first alternative of the definition of *likes*, confirming the previous type coercion of X : *person*, and merging coreference Y : T to X : *person*. The resolvent thus obtained is:

$$got(X : person, X).$$

This is not matched by anything in the database; and so, LOGIN must backtrack, reinstating the previous coercions and coreferences of resolvent (1).

As a next choice, *likes*(X : *person*, Y) unifies with the second alternative of the definition of *likes*, further coercing X : *person* to X : *peter*, and coercing Y : T to Y : *mary*. This produces the new resolvent:

$$got(X : peter, Y : mary).$$

This literal finds no match in the database; and so, LOGIN must backtrack again, reinstating the previous coercions of resolvent (1).

The third possible match is the last definition for the predicate *likes*, whereby Y : T is coerced to Y : *goodthing*. This yields the resolvent:

$$got(X : person, Y : goodthing).$$

For this, the only successful match is the third definition of the *got* predicate. This yields the empty resolvent, and the final result $X = mary$.

At this point, if forced to backtrack, LOGIN attempts the next alternative match for the initial invoking resolvent *happy*(X : T); namely, the second rule of the *happy* procedure. The next resolvent is thus:

$$likes(X : person, Y), \; got(Y, goodthing). \qquad (2)$$

A match with the first alternative of the *likes* definition merges *X* and *Y*. This gives the resolvent:

$$got(X : person, goodthing).$$

And this matches *got*(*mary*, *a*), producing the second result *X* = *mary*.

If backtracking is forced once again, resolvent (2) is restored. This time, as seen before, establishing the first literal of this resolvent eventually leads to coercing *X* : *person* to *X* : *peter*, and *Y* : T to *Y* : *mary*, resulting in the resolvent:

$$got(Y : mary, goodthing).$$

And this succeeds by matching *got*(*mary*, *a*).

Hence, this third alternative branch of computation succeeds with the final result *X* = *peter*. The reader is left to convince herself that there is no other solution for that particular query.

The next section illustrates a more complex example involving the presence of attributes.

6. A MORE COMPLEX EXAMPLE

The example of Section 5 was simple in the sense that it did not illustrate the use of inheritance among complex ψ-term objects - *e.g.*, records with attributes. One such example is next described.

In a type signature, such a type symbol as *person* has virtually any possible attribute typed as T. However, it may be desirable to constrain all possible database instances of *person* to be such that particular attribute types be more specific than T.

For example, let us suppose that we want to impose that every legal record instances of *person* in the database must have:

- a field *id* which must be a *name*;

- a field *dob* which must be a *date*; and,

- a field *ss#* which must have:

  - a field *first* which must be string of characters between '000' and '999';

  - a field *middle* which must be string of characters between '00' and '99'; and,

  - a field *last* which must be string of characters between '0000' and '9999'.

We can write this in LOGIN as:

```
person = (id => name;
          dob => date;
          ss# => (first => ['000'...'999'];
                  middle => ['00'...'99'];
                  last => ['0000'...'9999']))).
```

where, *name* is specified as, say:

```
name = (first => string;
        middle => string;
        last => string).
```

and *date* as:

```
date = (day => [1...31];
        month => [1...12];
        year => [1900...2000]).
```

The "$[\alpha \ldots \beta]$" notation is used to denote interval ranges over linearly ordered built-in types. For example, any string of character '*xy...z*' is an instance of the built-in type *string*, ordered lexicographically using, say, ASCII character codes. Thus, any interval of strings is a subtype of *string*; and unification corresponds to intersection. The same applies to types like *integer*, *real*, *etc.*

Now, let us suppose that we also want to specify that a *student* be a subtype of *person* - *i.e.*, that it inherits whatever attribute restrictions imposed on *person* - and that *student* further imposes restrictions on some attributes; *e.g.*, a *student* has a *major* which must be a *course*, and further constrains the *dob* field to be have a *year* between 1950 and 1970. This is achieved by:

```
student = person(major => course;
                 dob => (year => [1950...1970]).
```

Clearly, it must be checked that these type specifications are not inconsistent. And this can be done statically, before running a LOGIN program. Similarly, we could elaborate the rest of the signature. For example,

```
employee = person(position => jobtitle;
                  salary => integer).

workstudy < employee.
workstudy < student.

s1 = student(id => (first => 'John';
                    last => 'Doe');
             major => computerScience;
             ss# => (first => '897';
                     middle => '23';
                     last => '5876')).

w1 = workstudy(id => (first => 'Abebe';
                      middle => 'Nmougoudou';
                      last => 'Bekila');
               major => physicalEducation;
               ss# => (first => '999');
               salary => 10000).
```

Note that inheritance allows for ellipsis of information in the particular records of individuals in the database like *s1* and *w1*.

Now, we can define facts and rules in the context of this signature. For instance, part of a course enrollment relation could be:

```
takes(s1,[cs101,cs121,ma217]).
takes(w1,[pe999]).
```

To express that all students taking less then 3 courses are considered part-time students, we write:

```
parttime(X : student) :-
      takes(X,CL),length(CL,L), L<=3.
```

where length is trivially defined and computes the length of a list.

Finally, to formulate that all persons whose social security number starts with 999 is foreign, we write:

```
foreign(person(ss# => (first => '999'))).
```

Thus, a query asking for the last name of some foreign employee who is also a part-time student, and earns a salary less than 20000, is:

```
query(X : string) :-
      foreign(Y : employee(salary => Z)),
      parttime(Y : student(id => (last => X))),
      Z < 20000.
```

A remark worth making here is that extensive information can be processed statically for LOGIN before run-time. Thus, besides the inheritance type-checking already mentioned, some compile-time coercion must be performed to maintain consistent typing in clauses. Indeed, typing in the above query as we did would result in the automatic coercion of the ψ-term under the tag *Y*, transforming it internally into:

```
query(X : string) :-
      foreign(Y : workstudy(salary => Z;
                            id => (last => X))),
      parttime(Y),
      Z < 20000.
```

Thus, should $\perp$ occur in a clause by static type coercion, the clause would be eliminated.

We leave it as an exercise to the reader to verify that an answer to this query for the foregoing data is:

```
?- query(X).
X = 'Bekila'
```

7. CONCLUDING REMARKS

We have described of a semantically sound and operationally practical typed extension of Prolog, where type inheritance *à la* semantic network is cleanly and efficiently realized through a generalized unification process. We have illustrated how this can be achieved on some detailed examples. The language thus obtained is called LOGIN, an acronymic combination of "logic" and "inheritance".

The gain that we feel LOGIN provides over the conventional Prolog language is twofold:

1. the efficient use of taxonomic information, as well as complex objects;

2. a natural way of dealing efficiently with "set at a time" mode of computation essential to database applications.

In addition, we feel that the inheritance model behind LOGIN offers great potential for compile-time consistency checking, and object-oriented computation in a logic programming paradigm. For example, it is possible, at compile-time, to narrow drastically the range of indexing over a large database of individual records to only an appropriate view, based on the types of arguments of the predicates involved in the rules of a querying program.

REFERENCES

[ABN85] [AF82] [AITK84a] [ALN86] [ALN87] [AN86a] [AN86b] [AN86c] [AN87a] [AN87b] [BFL83] [CHUR40] [DK79] [FIND79] [HUET76] [KOWA78] [MINS75] [MM79] [REYN70] [ROBI65] [WARR83]

Chapter 10

Class Hierarchies in Information Systems: Sets, Types, or Prototypes?

Alexander Borgida
Rutgers University

ABSTRACT *One of the cornerstones of the conceptual modeling languages devised for the specification and implementation of Information Systems, is the idea of **objects grouped into classes**. I begin by reviewing the various roles played by this concept in the development and maintenance of Information Systems: specification of type information, repository of constraints to be verified, and maintanence of an associated set of objects (the "extent").*

Examples are then given to demonstrate that any realistic system in which data is to persist over a long period of time will need to allow occasional deviations from these constraints; i.e, classes specify prototypes. In consequence, I examine the problems of accommodating such persistent exceptions, and survey a mechanism which deals with these problems.

*I then consider a second feature of these languages -- the concept of **class hierarchies** -- and present arguments against the strict interpretation of class specialization and the notion of inheritance. Additional consideration of the concept of "default inheritance" leads to a list of desirable features for a language mechanism supporting non-strict taxonomies of classes: ones in which some class definitions may contradict portions of their superclass definitions, albeit in a controlled way.*

The effect of exceptions on the concept of "type correctness" of programs is considered in both of the above cases.

1. INTRODUCTION

I consider an Information System (IS) to be a computer system which maintains knowledge about some aspect of the world. In business data processing, such an IS usually consists of a large database of persistent facts, and a collection of application programs (transactions and queries) which are run against this database in order to retrieve or update information in it. It is by now conventional wisdom to consider an IS to be a *model* of the portion of the real world in which one is interested, and in databases this model is partitioned into two parts: (1) the *schema*, which captures generic, time-invariant information, and (2) the *facts*, which are specific, more volatile and occur in large quantities. The schema describes, among others, what I take to be the type system of the data in the IS[1], as well as constraints on legal data values (so-called "integrity constraints").

---

[1] The type system of the entire IS is obtained by combining the database schema and the transactions.

Conceptual Modeling languages used in Information System design are based on a few basic concepts:

- All information is recorded through the presence (or absence) of *objects / entities*, which are related to each other through *properties / attributes*.

- Similar objects are grouped together into *classes*, which prescribe the properties applicable to their instances, as well as constraints on the possible values for properties.

- Classes are organized into *subclass hierarchies* (also called IS-A hierarchies or taxonomies), where instances of a class A are also instances of its super-class B, and all constraints applicable to instances of B, also apply to instances of A.

For example, the classes of PERSONs, EMPLOYEEs and ADDRESSEs may be defined as in Figure 1.

```
class PERSON with
        name: STRING
        age:  1..120
        home: ADDRESS
        office: ADDRESS
        ...


class ADDRESS with
        street: STRING
        city: STRING
        state: {'AL,..., 'WV}
        phone: DIGITS(10)

class EMPLOYEE is-a PERSON with
        age : 16 .. 65
        degree : {'HSGD,'BSc,'MSc,'PhD}
        wages : 0.00 .. 80000.00
        manager : EMPLOYEE
        wageCap : (wages < manager.wages)
```

Figure 1-1: Some class definitions in a CML.

Here, EMPLOYEEs are defined as a subclass of PERSONs which have additional attributes such as jobCat and manager, or restrictions on attributes such as age; the attribute wageCap is a boolean-valued condition which must always be true for each object in EMPLOYEE. It is possible to view the attribute names as identifiers of constraints concerning instances of the class, so that every (class name, attribute name) pair uniquely identifies some constraint; for the definition of PERSON in Figure 1-1, one would get

```
class EMPLOYEE is-a PERSON with
        age : (self.age ∈ INTEGER) ∧ (16 ≤ self.age ≤ 65)
        ...
        manager : (self.manager ∈ EMPLOYEE)
        wageCap : (self.wages < self.manager.wages)
```

This ability to identify constraints by naming will be extremely useful when detecting and accommodating contradictions to constraints.

The presence at any one time of an object which is an instance of class EMPLOYEE is supposed to model the existence of some corresponding actual employee in the real world.

2. CLASSES IN INFORMATION SYSTEM DESIGN

Ignoring for the moment the subclass hierarchy, it is perhaps worth reviewing here some of the reasons for specifying the schema of an IS ahead of time.

From the **end-user's** point of view, the classes and attributes specified in the schema describe the "domain of discourse" of the IS: they delimit the area about which the IS is "knowledgeable", and define the vocabulary which one has to use to access this information. In addition, the constraints specified in class definitions help detect ubiquitous data entry errors, thus preserving the quality of the data, which is obviously in the interest of the users.

Classes provide a number of additional advantages for **implementors** of information systems, and these are categorized below under the various functions that classes play.

2.1. The class as type definition

One can consider classes simply as types, used to label the domains and ranges of operations in order to detect ill-defined operations on the data. For example, the above definitions indicated that wages, but not names, are numbers, which can therefore be used in arithmetic operations in transactions and queries. Also, classes viewed as types indicate the domains and ranges of attributes: wage is not applicable to arbitrary persons, only to employees. For this purpose, we might use the type system introduced by Luca Cardelli [CARD84]: it allows nested record definitions, and defines an appropriate notion of "subtype", which matches the notion of "is-a" subclass. For example, ADDRESS simply corresponds to the type of objects characterized by finite mappings from labels of fields (property names) to values in the property range:

$$[street : STRING, city : STRING, ...]$$

The utility of type systems and compile-time type checking for *detecting errors* in procedures is well known: For example, if we have a type inference/checking scheme, it is possible to detect situations such as attempting to evaluate the wages of an arbitrary person, who is not deducible to be an employee. Even if we decided this need not be an error, the compiler might insert run-time checks for this case, but avoid having to give them in provably correct cases.

2.2. The class as type identifier

It would be possible to enjoy some of the advantages of strong typing (e.g., knowing what attributes are applicable to an object) without associating identifiers with classes. For example, we could explicitly type the arguments of a procedure with the properties which we expect it to have:

```
procedure IncrementSalary(e :      [wages : INTEGER],
                          newSal : INTEGER)
    . . .
```

or we could do away with ADDRESS by expanding its definition "in-line":

```
class PERSON with
        . . .
      home:  [street: STRING
              city: STRING
              ... ]
      office: [street: STRING
              city: STRING
              ... ]
```

However, classes in fact represent correlated attribute structures, which are referred to frequently, and which are therefore assigned unique identifiers for ease of reference. Naming classes provides:

- a way of *abbreviating* definitions;

- a way of *localizing* definitions, which make changes easier, since we do not have to find all occurrences of the same type in the specification;

- the ability to *specify recursive definitions*, as in the case of employees, whose manager's are also employees.

The type systems of Pascal-like languages amalgamate the facilities to define types and to name them; we have separated them simply to emphasize that it is in fact possible to give type information without providing new type identifiers. Such a separation (present in languages such as Galileo [ACO85]) is desireable in the case when some type is only used in one location in the program, since this avoids cluttering up the schema with too many identifiers. For example, if for some reason office addresses should have additional attributes such as room# and companyName, then the definition of the new type could be done "in-line", as a modification of standard addresses:

```
class PERSON with
      home: ADDRESS
      office: ADDRESS [room# : 1..9999 ;
                       company : STRING]
          . . .
```

thereby avoiding the introduction of a new, infrequently used class: OFFICE_ADDRESS.[2]

---

[2] The semantics of our notation is that the specification room# : 1..9999 over-rides any potential definitions of room# in ADDRESS.

2.3. The class as a set of objects of a certain type

In many cases an *extent* is associated with a class, representing those objects which are instances of the class at some particular time. For example, EMPLOYEE would have as instances all objects of type EMPLOYEE currently in the IS, and their presence is supposed to model the existence of corresponding entities in the world. These objects are explicitly added or removed from the extent of this class by special operators *create/add* and *remove/destroy*.

Note that certain classes, such as INTEGER or ADDRESS, are usually considered to have an (uninteresting) fixed extent consisting of *all possible* objects of that type, rather than the set of all values of that type *currently appearing* in the IS. Furthermore, the equality of objects like addresses is decided on the basis of their attribute values, unlike entities such as persons, where object identity is an intrinsic property: two distinct persons may have identical property names. In some languages such classes are syntactically differentiated.

The extent is useful for a number of reasons:

- As we have stated, an IS is supposed to model the real world, which humans perceive in terms of entities which possess properties. It is therefore essential to record which entities exist and which do not, to locate entities based on their description, and to perform operations like counting entities. *Sets of objects* are the obvious data structure for this purpose, since one can *quantify* or *iterate* over them, as well as test membership in them.

- By creating classes corresponding to each discrete value of some attribute one can eliminate that attribute (e.g., jobCategory of EMPLOYEE can be replaced by various subclasses such as SECRETARY, DOCTOR,...).

- If the same selection condition occurs as part of many queries, it is convenient to be able to define the set of objects satisfying this condition, and thus allow the user to state the query more succinctly, stating only the additional restrictions required for selecting members of this special set. For example, if frequent reference is made to people with university degrees, one could define the class

```
class UNIVERSITY_GRADUATE is a EMPLOYEE such that (degree ≥ 'BSc)
```

2.4. The class as organizer of arbitrary constraints on relationships between objects.

When modeling any domain, one is trying to capture its semantics. If this semantics were to be expressed in predicate calculus, then classes would play the role of unary predicates (or possibly sorts, in a many-sorted logic), while attributes would be either functions or binary relations. There are many assertions in such a domain theory which do not seem to qualify as "type constraints". Some, such as limiting the age of persons to be between 0 and 120, or coreference constraints on attributes (e.g., x.spouse.spouse = x for a person x) appear to be like type constraints. However, arbitrary first order conditions involving the cardinality of extents (e.g., the average salary of EMPLOYEEs cannot exceed $45,000) or even uniqueness constraints (no two employees have the same Social Security number) seem to have little to do with type considerations, and are essentially tied to the extent of classes. Such constraints are however very useful since they

- detect errors in data being entered at run-time; and

- can be used at compile time to set up efficient storage and access structures (e.g., fixed-length record schemes, indices, B-trees based on keys).

What is common in all these cases is that such assertions can often be attached to one (or a few) classes. In this case, classes play an important role in *organizing the constraints*, and this organization is useful both during the design and the evolution of the IS.

2.5. Classes as objects

It is often convenient to view classes as objects themselves, so that they can be organized into meta-classes, and be assigned attributes of their own. For example, various subclasses such as SECRETARY, PROFESSOR, LIBRARIAN, etc. might all be made instances (not subclasses!) of the meta-class EMPLOYEE_CLASS, and each might have associated properties such as `avg_salary` (a property whose value might be obtained by summarizing over the extent of the class) and `avg_salary_limit` (which records some policy constraint of the organization). Note that such properties are clearly not attributes of individual employees.

Once we treat classes as objects, the property definitions of a class can also be viewed as part of its own attributes (e.g., in Taxis, the expression `EMPLOYEE..name` evaluates to STRING), thus making the schema of a database accessible to its users -- a desireable feature.

2.6. Discussion

I want to diverge slightly here and discuss briefly an issue which has arisen repeatedly recently: the need for associating extents with classes.

It seems quite clear that Information Systems must maintain at least one persistent set of objects: the set of all objects currently in the database. Without this, it is impossible to maintain a model of the world.

A very strong case can also be made for having *many* sets of objects. The extents of every class could in fact be recovered by adding an attribute to each of its instances, identifying the object as the instance of that extent. However, since an object may be an instance of many classes, one would need many such attributes or multiple-valued attributes, and queries or iterations over the set of objects corresponding to the extent of some currently existing class would then be more cumbersome to specify, and likely more inefficient to evaluate. In fact, Buneman and Atkinson [BA86] have argued convincingly that transactions need *additional* local, temporary sets of objects.

The next issue to consider is whether it might not be better to have two orthogonal concepts: *types/subtypes* and *sets of objects*. In the language Galileo [ACO85], it is possible to associate with each type one separately named extent. In practice, this is inconvenient since it appears to be an empirical fact that we want to associate extents with *most* entity types -- hence the need for many additional identifiers.

Extents could also be recovered if the predicates used in creating sets could involve tests on the type of objects. Buneman and Atkinson then suggest a clever way of avoiding the need for defining separate sets for each extent: a single function which takes a type as a parameter and returns the set of objects of that type. Such tests necessarily imply that types must be identified with their names, rather than their structure: otherwise sets

NEWSPAPER and JOURNAL could not be distinguished if they both had the same attribute structure:

```
[name : STRING, publisher : COMPANY]
```

This might be a problem in a language such as Amber [CARD85], where type equality is defined on the basis of *structure*. On the other hand, this procedure would probably not work properly for all possible types: for example, for the type INTEGER, would it return the set of all integers appearing anywhere in the database? What about integers used in the definition of the database? As a result, one will probably have to introduce types of types (e.g., ENTITY_TYPES) to restrict the domain of the FindExtent function. Furthermore, to avoid inefficiency, this function would probably have to be implemented differently: traversing a predefined chain of objects, rather than going through all óbjects, and testing each individually for membership in some type. In fact, this is likely to be a predefined function -- i.e., the concept of extent comes predefined with each type, just as the extent of classes, which was my point earlier. The difference between the two approaches then seems to come down to one of syntax: differentiating the type name (EMPLOYEE), from the *automatically associated* set (EMPLOYEE^), and this may well be a useful distinction to make in the syntax.

Buneman and Atkinson are however correct that one needs temporary sets in transactions, which can either be accomplished by having set types in the language, or through dynamically created *classes*: i.e., local class constants, specified as instances of metaclasses. The language DIAL ([HB80]) in fact utilizes this idea to associate temporary attributes with objects as part of a transaction. The full details of such an approach are yet to be specified however.

A final problem with extents arises in the case of multiple databases[3] : both Rutgers' and AT&T's database may have a class of EMPLOYEEs, and if these are merged, obvious confusion arises about the extents of the same class. The problem of multiple databases cannot however be resolved by simply having two separate sets associated with the class EMPLOYEE: attribute names also need to be dissambiguated (e.g., is this the AT&T salary of an employee, or his Rutgers salary?).

A general solution is to allow all identifiers, like class names and attribute names, to be indexed by the *database* with respect to which they are to be evaluated, and then provide a conventional abbreviation of the form **with** *<database>* **do**, (similar to the Pascal *with* *<record> do*), which allows programmers to omit the explicit reference to the database:

```
with at&t-db do ...    for x in EMPLOYEE do if x.salary  ...
```

3. EXCEPTIONAL CLASS INSTANCES

Unfortunately, the framework for IS design presented above runs into problems when faced with reality: In most cases, ISs deal with domains in our everyday world, domains populated by concepts known to philosophers as "*natural kinds*"; there is considerable evidence that, unlike concepts such as "triangle" or "stack", natural kinds like "employee" or "chair" do not have clear-cut definitions, and our understanding of them is often based on complex notions of "prototypicality". In particular, facts accumulated about a natural domain over some reasonable period of time seem incompatible with a *predefined, static*

---

[3] This problem was brought to my attention by Alan Snyder and David Maier.

type scheme which is strict, in the sense of not admitting any deviations. I will try to illustrate this with a number of examples taken from the context of an IS which holds information about employees, their salaries, degrees, addresses, etc.

- The employees have associated addresses, and at the time the system is designed all employees live in the USA. It seems reasonable to describe as part of the schema the type ADDRESS of addresses in the USA. But suppose that after a few years, the company hires a single employee in France. Her address is of a different type, and would not fit into the database. Furthermore, there seems to be no reasonable way to anticipate such situations: this exceptional employee may have been sent to any country, and although it is possible to describe the type FRENCH_ADDRESS, I claim that defining the non-trivial type POSTAL_ADDRESS of all addresses in the world is beyond the abilities of the average IS developer.

- Continuing with the example of the foreign employee, her contract calls for a salary of FF 450,000. Storing the number 450,000 in the wages field, which normally holds US$, is inappropriate since it would invalidate programs calculating budgets for example. On the other hand, one cannot store an equivalent US$ value because of the fluctuating exchange rate.

- As a final example of type violation for record attributes, consider degree: On the rare occasion when a foreign-educated employee is hired, with a French *Baccalaureate* for example, one is again forced to either omit the information from the database or "trim it" to fit the existing mold, thereby losing information since a "Bac" has no American equivalent.

- Occasionally, one may want to store a new attribute for some employee (e.g., a security clearance code) or make a normally single-valued attribute (e.g., spouse) be set or sequence-valued (e.g., for a polygamist).

- Finally, almost all constraints on legal values (e.g., "employees earn less than their managers") are subject to exceptions.

I wish to draw several conclusions from the wide variability of facts in the world, and the pervasiveness of special cases:

1. In designing IS software, it is important to be able to abstract away the rare, special occurrences at first, in order to concentrate on the normal cases. Like other forms of abstraction, this one allows designers to cope with the multitude of details that is the bane of IS design.

2. Those special cases which are *anticipated* should be added as "annotations" of the main program text, rather than embedded in it, because this leads to more readable programs, which are therefore easier to maintain.

3. There will always be rare, unanticipated cases, and one must be prepared to distinguish these and accommodate them at run-time.

Consider some of the problems which arise when trying to accommodate exceptional facts. At least some type systems would appear to have no difficulty in assigning a type to an arbitrary new specific fact, such as a French address, and could automatically extend a type such as ADDRESS through type union to cover this special case. Unfortunately, the transactions cannot be similarly extended in an automatic way: some of them will be incorrectly typed after this extension, others just semantically incorrect (e.g., the French salary). The obvious approach of modifying all such programs before resuming computation has several drawbacks:

- It may require a great deal of effort to be spent on recompilation.

- Some of the recompilation effort, but especially the programming effort of redefining the procedures, might be wasted if some transaction is invoked only after the exceptional data has been updated (e.g., the employee with a French degree is fired before he comes up for promotion).

- Current data storage and retrieval techniques may result in quite inefficient storage schemes because they ignore the rarity of the special values. For example, space may be allocated for the longest possible name, as opposed to the longest likely name, or Social Security number may not be used as a key because of the rare occurrence of duplicates (thus requiring a new unique identifier field).

In [BORG85], I present instead an essentially "demand driven" technique for dealing with exceptional ("atypical") values in interactive systems. It tries to combine the benefits of compilation for the normal or anticipated special cases, with those of run-time interpretation for unanticipated abnormalities. The following are some of the chief features of this approach:

- The exception handling mechanism of the programming language allows transactions which signal or propagate exceptions either to be "backtracked", so that all side effects are undone, or to be "resumed".

- The exception handling facility of the programming language is adapted to play a dual role:

  1. to detect and react to violations of constraints, including type constraints;

  2. to detect and react to the encounter of exceptional data, which has been marked exceptional when the initial violation of some constraint was **resumed**.

- Exceptions propagated to the top level, and not handled by "default handlers" set up ahead of time, are handled by procedures provided by users of the IS "on-line". As part of such handlers, users can extend the taxonomy of markers used to distinguish various kinds of exceptional data.

The following brief scenarios illustrate our mechanism in action:

Suppose a user-invoked operation has the effect of assigning marx a salary of 95000:

```
marx.wages := 95000
```

The update operation detects this as a violation of one of the constraints stated on EMPLOYEE, and therefore raises an exception.

The user may realize that this was an error, and reinvoke the transaction with corrected arguments. Or, she may decide that the value is correct in which case she would **resume** the transaction (and the update operation), which would result in the value being stored. But before this, she would have to create an object which indicates that marx's wages are exceptional.

If later, a transaction tries to retrieve and use marx's salary in a statement like

```
y := x.wages
```

the program is again interrupted by an exception (the object marking the property exceptional is used to signal the exception), which indicates that something unusual is about to happen.

The exception handler invoked in this case may abandon the transaction as inapplicable in this case. Or, the handler may **resume** it, thereby retrieving the value and continuing the computation with it. For convenience, we also offer a **resume with** <value> command, which allows the exceptional value to be replaced by another one *for this particular context* (e.g., using BSc for Baccalaureate).

Note that a default handler may be associated with the object marking an exceptional property, and this may invoked if the user does not wish to provide a special one for this situation.

3.1. Exceptional instances and type safety

By allowing *interpreted* procedures to replace calls to *compiled* ones at various levels in the invocation hierarchy, we are basically allowing users of the system to provide variants of the original transactions which are tailored to fit the specific data encountered so far, including the exceptional values. The price payed for this is the inability to do "strong typing" of the programs: procedures can be type-checked as before under the assumption that they deal with unexceptional values, but once a program is resumed after an exception, all values become unreliable, and only run-time checks can ensure that errors are avoided. It is worth noting that exceptional values, marked as suggested above, could be used even with a model of exception handling where interrupted procedures are terminated or backtracked (and never resumed); in this framework one could easily adopt the scheme presented by Cardelli in [CARD85], where so-called "dynamic" values carry type-information with them at run-time and these can be explicitly checked by the programmer. The effect would be to allow exceptional values to be dealt with in a "type safe" way, in the sense used in languages like ML. In the resulting language though, every variation of a procedure dealing with special values would have to be completely rewritten, which is certainly more inconvenient.

I also wish to point out that the problems posed for type safety by exceptional values appear to be similar to those caused by uninitialized variables in any programming language, and by so-called *null values* used to indicate lack of information in ISs. In fact, in [BORG85] I show how various kinds of null values can be handled within the framework of the proposed exception handling mechanism.

4. GENERALIZATION HIERARCHIES AND INHERITANCE

Consider an IS for use in a hospital, which keeps track of its employees, doctors, patients, etc. The schema presented earlier could be extended by the definitions in Figure 2, together with numerous other subclasses of patients (grouped by disease), doctors (grouped by speciality), etc.

Hierarchies of classes are in fact evidence of a form of abstraction at work: *generalization* suggests that one abstract away the detailed differences of several class descriptions, and present the commonalities factored out as a more general superclass. Thus PATIENT and PHYSICIAN are generalizations of various kinds of patients and doctors respectively, and PERSON itself is an abstraction of PATIENT, PHYSICIAN, EMPLOYEE, etc. Specialization, the refinement process corresponding to generalization,

```
class HOSPITAL with
         name : STRING
         location : ADDRESS
         accreditation: {'Local, 'State, 'Federal}
         ...


class PHYSICIAN is a PERSON with
         affiliatedWith: HOSPITAL
         certifiedBy: {'AMA, 'ABFP, 'ABS,...}
         ...


class PATIENT is a PERSON with
         treatedAt: HOSPITAL
         treatedBy: PHYSICIAN
         ward : WARDS
         ...


class CANCER_PATIENT is a PATIENT with
         treatedBy: ONCOLOGIST
         ...
```

Figure 2: Some additional subclasses for a hospital IS.

then allows one to introduce details by choosing subclasses of existing classes and specifying for each the additional information characterizing it: this includes additional properties applicable to its instances, and additional constraints both on the new and old properties. Note that attributes and constraints stated on the superclass need not be restated for the subclasses: they are assumed to hold by *inheritance*. Thus patients and doctors also have names, home and office addresses, etc. which are inherited from PERSON. Note that during specialization, the range of an attribute (e.g., treatedBy) can be modified for a subclass (e.g., CANCER_PATIENT), as long as the new range (ONCOLOGIST) is a proper subclass of the original range (PHYSICIAN).

The result is a taxonomy of classes, organized in a partial order. Such taxonomies can then perform at least one software engineering task during the development of a knowledge base: by considering class definitions starting from the top of the hierarchy, we are making use of an abstraction principle that allows details to be introduced gradually and systematically.

The additional advantages of class hierarchies can be reviewed and summarized using the categorization of the various functions of classes introduced earlier:

4.1. Subclasses and types

If classes are used to type the arguments of procedures which manipulate the data, then the concept of supertype is useful since it allows a form of "polymorphism": a procedure whose argument has type PERSON, will work when given an argument which has as type any subclass of PERSON, thereby avoiding the need for defining separate versions of the same procedure. Note that because of the over-loaded meaning of "class", in some cases it is more appropriate though to interpret the procedure definition P(e:C) as a constraint requiring P to be applied only to instance of the extent of C.

4.2. Subclasses and type identifiers

As before, the chief advantages of this view appear during the evolution of the software:

- class descriptions are abbreviated, hence more readable;

- there is less chance for inconsistencies to creep in, because the same fact is no longer repeated in several places;

- during IS evolution, changes in a class at a higher level are known to be propagated uniformly to all subclasses; thus errors of omission are avoided by localizing changes.

As before, it is however convenient to allow more specialized types to be introduced during specialization without requiring names to be associated with them if they are to be used only once: For example, if all we cared about oncologists is that they are physicians certified by the American Board of Oncologists (ABO), and oncologists only appear in the definition of cancer patients, then this class could be defined as

```
class CANCER_PATIENT is a PATIENT with
    ...
    treatedBy: PHYSICIAN [certifiedBy: {'ABO'}]
```

This would reduce the size of the schema by avoiding the addition of a new class and name, and thus reduce the (conceptual) clutter for users and designers.

4.3. Subclasses and extents

The class hierarchy expresses subset constraints on various extents: the set of all ALCOHOLIC is a subset of the set of PATIENTs[4]. An important advantage of this constraint on the subclass relation is that it allows the extents of classes to be manipulated in a much more concise way: if an object is added to the extent of ALCOHOLIC, it is automatically added to the extents of all its superclasses, including PATIENT, PERSON, etc. If the extent of classes was replaced by sets, then one would need to write for every class separate procedures for adding or removing objects from its extent in order to ensure that the appropriate subset relationships would be maintained; these procedures would unfortunately be likely to become error prone as the class hierarchy evolves. Of course, declaring the subset constraints separately and having the system maintain them is a new language feature, and thus has no discernible advantage over the "class with extent" approach.

4.4. Subclasses and the organization of constraints

Since subclasses are supposed to describe subsets of elements of superclasses, the definition of a subclass must be consistent with that of its superclasses. Thus the age

---

[4] Note that in a type hierarchy with exceptions, the type ALCOHOLIC may not be a subtype of type PATIENT, even though the extents are required to be subsets.

restrictions of EMPLOYEEs must imply the age restrictions of PERSONs. Both editing tools and compilers can use this as a consistency check for detecting errors in class definitions, similar in intent to type-checking of procedure arguments with parameter specifications.

Note that many of the above advantages are built on the assumption that subclass specifications do not contradict those of superclasses. Note also that all the advantages of class hierarchies and inheritance mentioned above are strictly ones of convenience and software engineering: these concepts can be eliminated from any IS without changing in any way the facts that can be stored. Therefore these same criteria will play a significant role in the arguments below concerning the exact nature of class hierarchies.

5. NON-STRICT CLASS HIERARCHIES

5.1. Problems with strict inheritance

We have argued earlier that in understanding and describing real-world situations it is often natural and convenient to ignore at first rare or exceptional cases, concentrating on the usual or normal cases. Unfortunately, in some cases this abstraction applies naturally together with the generalization abstraction, resulting in *over-generalizations*: a paradigmatic example of this is the commonly held beliefs that "birds fly", at the same time as "penguins and ostriches are birds" yet "penguins are flightless" and "ostriches are flightless". We shall illustrate a number of such problems arising in the domain of the hypothetical knowledge base for the hospital.

The simplest and most common form of contradiction arises when the range of an attribute for a specialized class is not necessarily a subclass of the corresponding range for the superclass. For example, one may want to define ALCOHOLICS to be patients who, among others, are treated by psychologists:

```
class ALCOHOLIC is a PATIENT with
       treatedBy: PSYCHOLOGIST
       . . .
```

But psychologists usually are not physicians, and therefore do not have the same applicable attributes and constraints. Therefore the above is not a proper specialization of PATIENT. Note that the intended interpretation of this definition is not that alcoholics are treated by persons who are both physicians and psychologist (in which case there would be no contradiction) -- they need only be psychologists.

A similar problem arises when an attribute becomes inapplicable to all the instances of some subclass: for example, although we may want to record a ward for PATIENTs, this attribute would be inapplicable to AMBULATORY_PATIENTs. Although one could consider this as specializing the range of ward to the empty set, it is I believe more appropriate to state that this attribute is incorrectly applied to such patients:

```
class AMBULATORY_PATIENT is a PATIENT with
       . . .
       ward : NONE
```

where it would be a type error to attempt to evaluate the ward of a such a patient.

150

Similarly, when a modification of some type is defined as part of the range specification of a refined attribute, it is occasionaly convenient to allow the new type not to be a proper subtype of the original one: for example, if we required tubercular patients to be treated in some specific foreign country like Switzerland, then these would not have accreditation, and their addresses would be modified as follows:

```
class TUBERCULAR_PATIENT is a PATIENT with
     ...
     treatedAt: HOSPITAL [accreditation : WRONG;
                          location: ADDRESS [state : WRONG;
                                             country : {Switzerland}]
                         ]
```

Such addresses are not proper subclasses of the class ADDRESS, hence these hospitals are not a subtype of HOSPITAL, and therefore TUBERCULAR_PATIENT is not a proper specialization of PATIENT.

5.2. Alternatives to non-strict inheritance

5.2.1. Strict inheritance with reconciliation

The most obvious solution is to generalize the portion of superclass description which is being contradicted: PATIENT° could be treated by HEALTH_PROFESSIONALS, which would have as subclasses PHYSICIANS and PSYCHOLOGISTS; this would allow ALCOHOLICS, defined as before, to be a proper specialization of PATIENT°. Most other kinds of patients would however be treated only by physicians, so one would have to laboriously specialize the treatedBy attribute for CARDIAC, CANCER, etc. patients to specify in each case PHYSICIAN as a range. This essentially negates one of the significant advantages of inheritance: the factoring out of commonalities which need not be repeated.

5.2.2. Strict inheritance with intermediate classes

To recapture the advantages of inheritance, one could introduce intermediate classes whose only role is to act as anchors for inheritance. So for example, PATIENT_TREATED_BY_PHYSICIAN could be specified as a subclass of PATIENT°, and then specialized to obtain CARDIAC, CANCER, etc. patients. The first disadvantage of this solution is that it *clutters up* the IS with definitions of dubious utility, thereby making it harder for users to find the useful classes they are seeking. Also, exactly such classes are likely to have uninteresting extents. The second disadvantage is that of *combinatorial explosion*: Suppose some class C has two attributes p and q

```
class C with
      p:D
      q:E
```

which need to be generalized in order to avoid contradictions with a couple of exceptional subclasses. This would result in a new class C°, but we would need to define three specializations of it: one in which p is again restricted to D, one in which q is restricted to E, and finally one in which both restrictions apply. Furthermore, every time we wanted to create some new subclass of C, we would have to decide which of the four classes C°,..., C_4 should actually be specialized.

5.2.3. Default inheritance

A popular approach in Artificial Intelligence is to adopt the convention that the "closest" constraint in the hierarchy overrides all others, including ones that are contradicted. For example, if the taxonomy is a tree, then one can say that the assertion on the nearest ancestor in the taxonomy holds, and hence the inherited property can be computed efficiently by searching up the subclass tree. "Default" approaches face several problems as part of a definitional mechanism.

The first problem, noted already elsewhere, is that the search-based definition is no longer well-defined once the classes are organized in a partial order, not just a tree: if class A has two ancestors, B and C, both of these could specify constraints on A by inheritance, and it is not specified which one should be chosen.

A more significant problem, from our point of view, is the fact that it is no longer possible to detect inconsistent definitions because the system cannot distinguish erroneous definitions from defaults: Whenever the definition of some class contradicts an assertion on one of its superclasses, the contradiction could be intentional or accidental, and it is impossible to tell which is the case without further user intervention.[5] For the same reason, when adding a class definition somewhere in the middle of the IS-A hierarchy or when modifying a definition, it is impossible to be sure that the constraints will be inherited by all subconcepts or whether some intervening concept will accidentally block the inheritance. In fact, in all languages which have "cancellable inheritance", one can find out if some property of a class is universally true only by checking all of its subclasses.[6] Significantly, these problem arise especially when a large IS is being modified by someone else than the original designers.

Along different lines, default inheritance is usually defined only for subclass hierarchies so that it is impossible for the definition of a class to supplant some constraint on one of its attributes (e.g., the `treatedBy` attribute of TUBERCULAR_PATIENTS) without defining a special subclass of these attribute values (HOSPITALS_IN_SWITZERLAND and SWISS_ADDRESS) for which the constraint in question can be cancelled. This leads once again to the proliferation of unnecessary classes.

Finally, there appear to be considerable difficulties in providing clear and simple semantics for "default inheritance" when the hierarchies are not trees.

5.3. Desiderata for non-strict class hierarchies

To summarize, any mechanism for dealing with non-strict specialization should have the following properties:

- inheritance: retain the abbreviatory advantages of inheritance, and the advantages it provides in localizing modifications;

- veracity: allow the redesigner to decide when exactly a constraint will hold, without having to search blindly for contradictions;

---

[5] The situation is analogous to that which arises in programming languages which do not have typed variables: if x originally held a string, and y is integer valued, then is "x:=y" a typo (having forgotten the quotes around y) or intentional?

[6] Brachman has argued this case more extensively and cogently.

- locality: allow incremental changes to be made locally, without having to modify earlier definitions;

- semantics: have a clear semantics, especially for the case of non-tree hierarchies.

In a forthcoming paper we present a solution which claims to meet these criteria. It is based on the idea of *explicitly acknowledging* contradictions between various specifications, and using such explicit "excuses" to reconcile the constraints of various classes. We content ourselves with presenting in Figure 3 the syntax of the proposed solutions. For example, it is specified that the constraint identified by the pair (PATIENT, treatedBy) for some object <u>self</u>, is excused from holding in the case when <u>self</u> is an instance of ALCOHOLICs, and the constraint (ALCOHOLIC, treatedBy) is in force.[7] Note that the semantics continues to require that the extent of ALCOHOLIC be a subset of the extent of PATIENT.

```
class ALCOHOLIC is a PATIENT with
     treatedBy: PSYCHOLOGIST excuses treatedBy on PATIENT
     ...



class TUBERCULAR_PATIENT is a PATIENT with ...
     treatedAt: HOSPITAL
                [accreditation : WRONG
                    excuses accreditation on HOSPITAL ;
          location: ADDRESS
                    [state : WRONG
                        excuses state on ADDRESS ; ]
                    country : {'Switzerland}
                    ]
                ]
```

Figure 5-3: Syntax of defining subclasses which are not subtypes.

5.4. Non-strict inheritance and type inference

Consider the following procedure for deciding if a patient is treated by a doctor in his home town:

```
procedure  IsLocallyTreated(p:PATIENT) returns BOOLEAN
     begin
       if p.home.city = p.treatedBy.office.city
           then return(true)
           else return(false)
     end
```

If we assumes that the class hierarchy was strict, then this procedure would have no type errors, given the definitions of PATIENT and ADDRESS. If we allowed exceptions in

---

[7] Observe here also how the ability to identify constraints by an attribute and class name is crucial in expressing the desired conditions

the class hierarchy (e.g., TUBERCULAR_PATIENTs), then the above procedure happens to still be type-safe, but not if we replaced `city` by `state`. This difficulty in checking the type-correctness of procedures is the result of several factors:

1. The restriction `p:PATIENT` can more properly be interpreted as a condition on the membership of p in the extent of PATIENT: i.e., p may have the properties specified in the declaration of PATIENT or its subclasses.

2. Viewing `p:PATIENT` as a type constraint is too strong for the purposes of this procedure. Using the syntax and semantics of types in [CARD84], it is sufficient to have p be of type

```
[home : [city : STRING];  treatedBy : [office : [city : STRING]]]
```

and ensure that all instances of PATIENT satisfy this constraints. The difficulty here is that it is evidently too cumbersome for programmers to specify such types, and much easier to use the identifier of some already existing class, even if it comes with additional "baggage".

A type inference or checking system is the obvious tool to use to derive the types of procedure parameters, and then verify that indeed all instances of the given class have this more general type. The exact nature of the type system necessary for this task is the subject of a forthcoming paper.

6. SUMMARY

I have tried to accomplish several goals in this paper. First, I distinguished a number of different uses for *classes* in Conceptual Modeling languages for Information System design. These included:

*   presenting *type* information about the objects in the database, especially the attribute structure of objects, as well as the transactions and queries manipulating them;

*   describing *sets* of objects which are useful in defining iterations or quantified expressions;

*   *organizing* the definition of the IS, especially the consistency constraints.

The ability to name classes, group them in subclass hierarchies, and the mechanism of inheritance then allowed the description of the IS to be abbreviated and localized, thus making the software more understandable and modifiable.

Although the role of a class as a type definition and as specifier of an extent are clearly distinguishable, I have argued that they empirically co-occur in the context of the usual data-processing Information Systems, and that there seems to be no obvious way of obtaining one of these aspects from the other.

The main thesis of the paper is however that in IS programs, it is difficult and impractical to consider a class as defining *strictly* the type of its instances. Two reasons were advanced for viewing a class definition as the description of a frequently occurring *prototypical* instance:

1. Because of the variability of the natural world, and because an Information System persists over a long period of time, for every constraint there is some conceivable situation where it could be violated. However it appears to be impractical, if not impossible, to anticipate all such cases. By permitting atypical class instances to co-exist with the normal ones, we ensure the completeness of the Information System without eliminating all controls over its accuracy, or losing all information about the structure of the instances.

2. Since prototypical instances of a class (e.g., PATIENT) need not be prototypical of its superclass (e.g., PERSON), one can expect overgeneralizations to occur even during design: a class description may occasionally be contradicted by one of its specializations. Furthermore, permitting such contradictions provides many of the same advantages as those conferred by inheritance.

In both of the above cases, *deferment of special cases* provides an important technique for organizing the IS software, both during design and maintenance.

Note that the need for an explicit extent, as opposed to a set obtained by testing objects for their type, is even clearer when exceptional instances are allowed into the database, since these cannot be considered to meet the type specification of their class.

Although the main purpose of this paper was to defend the above theses, I have also summarized a technique, more fully described elsewhere [BORG85], for accommodating exceptional class instances, presented the desiderata of a language mechanism for dealing with contradictory class definitions, and suggested that the view of classes as prototypes need not negate all the benefits of type checking.

7. ACKNOWLEDGEMENTS

I have been steered to consider the relationship of classes to types by very useful discussions with Peter Buneman, David MacQueen and John Mitchell. The faults in the analyses are of course my own. This research was supported by funds from Rutgers CAIP/NJCT and NSF-MCS-82-10193.

REFERENCES

[ACO85] [BA86] [BORG85] [CARD84] [CARD85] [HB80]

Chapter 11

Language and Methodology for Object-Oriented Database Environments[1]

Stanley B. Zdonik
Peter Wegner
Brown University

ABSTRACT *This paper describes an object-oriented database language being implemented at Brown for use on workstations, and demonstrates its use in defining an object-oriented programming environment. The database language is illustrated by specifications of the UNIX file system, Ada packages, program structure, and multiple views of program modules. Each example illustrates a different feature of object-oriented programming methodology. Collectively the examples serve both as an introduction to our database language and as a tutorial for object-oriented system programming.*

1. INTRODUCTION

The object-oriented database environment described in this paper integrates ideas from the programming language, database, and environment worlds. The language approach builds on the ideas of Simula, and Smalltalk [GR83], on multi-paradigm languages such as [ACO83] and [BS83], and on object-oriented models of type and multiple inheritance [CW85] [CARD85]. The database approach follows the tradition of semantic data models [CM84] [HM81] [MBW80] [ZDON84]. In the environment area we have been influenced by the graphical approach of Reiss [REIS84,REISS] as exemplified by the Pecan and Garden systems.

The set of research issues that must be addressed indeveloping an object-oriented language, environment, and methodology is much larger than can be dealt with in this brief paper. We here restrict ourselves to a small subset of potential topics: namely the description of our object-oriented database language, and the methodology of using this language for object-oriented system programming. We do not address implementation issues such as garbage collection. This allows us to focus on issues such as the relation between language and database features in an object-oriented language, the granularity of objects in object-oriented compilers, and the inheritance of system attributes by application modules.

The system programming domain is chosen to demonstrate that our object-oriented language may be used not only for traditional applications in data processing but also for less traditional applications, such as environment building. However it should be emphasized that the database language is general purpose and that its domain of application is in no way restricted to system programming. A broader overview of the research issues we are currently exploring is briefly given in the appendix.

---

1 This work was supported in part by IBM Yorktown Heights and in part by ONR under Contract N00014-83-K-0146 and DARPA under order No: 4786.

The architecture of our programming environment is based on a database that contains all objects used for progamming in the large, including such things as modules, compilers, libraries, editors, and debuggers. It is possible to express relationships between these object such as *imports-from, exports-to, inherits-from, is-an-instance-of, is-an-interface-of,* and *is-a-version-of.* The database provides the glue that integrates these pieces. If the database is powerful enough to describe the semantics of the environment, it can be responsible for enforcing some of the common constraints. The constraints do not have to be reprogrammed within the contexts of each tool in the environment. This makes constructing new environments much easier.

We have introduced elsewhere [ZW85] the notion of a "databased programming language" as a programming language with database features sufficiently powerful to describe its own environment. Our object-oriented language is an example of a databased programming language. It includes the following features:

- Abstract data types with inheritance

- Classes - aggregates with associative retrieval

- System database types from which persistence and version control is inherited

- Operations, properties, and types which are themselves database objects Objects which have multiple interfaces.

- Triggers, which can be used to coordinate activities.

The first of these features by itself is characteristic of object-oriented languages like Simula or Smalltalk. The remaining features provide database facilities for organizing and manipulating collections of persistent objects. Thus an object-oriented language with just the first feature becomes an object-oriented database language by virtue of the fact that objects have the remaining properties.

Section 2 below introduces each of the above features using a file system specification as an illustrative example. The remaining sections illustrate the methodology of object-oriented environment specification in greater depth by defining Ada packages, Pascal program structures, and multiple views of programs as collections of objects in the database.

2. LANGUAGE AND DATABASE FEATURES

2.1. Object Types

Object-oriented programming is based on the specification of object types by a collection of operations and properties and by supertypes from which additional operations and properties may be inherited.

```
Define Type T
  Inherits from: <list of supertypes>
  Operations: <list of operations>
  Properties: <list of properties>
```

This kind of type definition is powerful and flexible. It captures data abstraction as a special case when there are only operations and no properties. It captures records as a special case when there are only properties and no operations. The inheritance mechanism allows the new type to be defined incrementally by specifying only those operations and properties in which the new type differs from previously defined types. Since operations may be viewed as special kinds of properties and properties may be specified by a pair of get and put operations, we could specify object types in terms of just a set of operations or just a set of properties. But there are advantages in expressive power in allowing both operations and properties in specifying object types.

The representation of a type can be accessed only by the operations that are defined on that type. Operations of a type can access instances of other types solely by means of the operations of that type. An operation of a subtype S can only interact with a supertype T by means of the exported interface (i.e., operations and properties) of T.

2.2. A File Example

The following definition of "*File*" has a supertype "*Entity*", three operations, and one property:

```
Define Type File
  Inherits from: Entity
  Operations:
    Create-file ( ) returns (F:File)
    Open (F:File) returns (F:File)
    Close (F:File) returns (F:File)
  Properties:
    Filename: String
```

Once the basic notion of a file has been defined, we may define particular classes of files that specialize this general notion by object types that inherit the above file properties. A file directory (similar to that used in UNIX [BOUR78]) might be defined as follows:

```
Define Type Directory
  Inherits from: File
  Operations:
    Enter-file (D:Directory, F:File) returns (D:Directory)
```

Here, the *Directory* type inherits all the properties and operations from the type *File* since *File* is defined to be a supertype. A type can have multiple supertypes and can thereby inherit properties and operations from several types by multiple inheritance.

The object-oriented paradigm is well suited to certain kinds of type changes. It supports the creation of new types that add semantics (i.e., operations or properties) to pre-existing types. The new type can be simply added to the appropriate place in the type hierarchy. For example, typed files can be defined by simply adding operations for typed objects to the previous type definition. Thus text files with an add-character operation and record files with an add-record operation may be defined as follows:

```
Define Type Text-File
  Inherits from: File
  Operations:
    Add-Character (F:Text-File, C:Char, At:Integer)
```

```
Define Type Record-File
  Inherits from: File
  Operations:
    Add-Record (F:Record-File, R:Record, At:Integer)
```

2.3. Classes and Associative Retrieval

The ability to define objects in terms of their operations and properties and to define new types in terms of the way they differ from existing types is a property of object-oriented languages. We enrich this linguistic base by database features that transform the language into a tool for specifying object-oriented databases. In doing this we find that the object-oriented features at the language level are very useful in providing a framework for the introduction of database features.

Objects may have an associated *class*[2] that

contains all instances of the type. Classes are collections of objects of uniform type similar in their properties to flexible arrays or tables in other languages. Whenever a new instance is created, it is automatically inserted in the appropriate classes. Operations on classes include associative retrieval of objects by specification of values of a subset of the properties. A more complete specification of object types that includes a class specification is as follows:

```
Define Type T
  Class: <class-name>
  Inherits from: <list of supertypes>
  Operations: <list of operations>
  Properties: <list of properties>
```

Classes support associative retrieval of a subset of members satisfying a predicate by means of a "select" operation. The following select operation selects all elements of a class C satisfying a predicate P and forms a new class S containing the selected elements:

```
Select (C:Class, P:Predicate) returns (S:Class)
```

The predicate P is constructed by using standard query language capabilities.

Classes may be formed by accumulation of elements of a type, by selection of members satisfying a predicate, by logical union and intersection operations and by explicit enumeration of members. They provide a mechanism for expressing database constraints.

If no class is explicitly mentioned in a type definition, we assume that a class with the same name as the type is automatically created.

---

2   We borrow the term *class* from the database literature and systems such as the SDM [HM]. In the SDM a class is a collection of objects of a homogeneous type. What SMALLTALK calls a class is really closer to the programming language notion of a type, and there is a growing literature on this topic [CAR85b].

2.4. System Types for Persistence, and Version Control

The objects in the database are persistent by virtue of the fact that the highest-level type, the entity, has system defined attributes of persistence. Entities also have implicit locks that prevent concurrent write access by more than one user, and provide concurrency control over all objects at the database level. Version control is provided by a type called *History-Bearing-Entity* (HBE). Any new type can inherit the behavior of the version control mechanism by being defined to be a subtype of HBE. In addition to this system-defined version control mechanism the user (or system administrator) could define a different version control mechanism that could be inherited in place of the standard mechanism for an application with special version control requirements.

Basic database attributes can be provided in an object-oriented, easily modifiable way as attributes of high-level database objects. This allows the definition of such attributes to be localized but at the same time widely shared by inheritance.

2.5. Operations, Properties, and Types are Database Objects

By making operations, properties and types objects of the database we achieve great expressive power without sacrificing uniformity or proliferating language mechanisms. Persistence is automatically inherited by operations, properties, and types, while version control may be optionally inherited. Database facilities may be used for establishing useful behavior for objects, such as the class mechanism or interfaces, and triggers may be used to establish desirable side-effects to operations.

Since properties have types, they can have arbitrary properties and operations of their own. The most general property type, type *property*, supports two special operations *get-property-value* and *set-property-value*. All propoerty types inherit these operations. Values of properties can be computed dynamically by the specialized code for the *get-property-value* operation.

By viewing each type as an instance of the type *Type*, we can use our version control mechanism to keep track of arbitrary changes to types themselves. In this way, type *Type* is defined as a subtype of HBE. Note however that the fact that the type *Type* is itself a type does not lead to a paradox because types are not interpreted as sets but as templates.

2.6. Interfaces and Triggers

Interfaces and triggers are two special kinds of relations among objects that are provided because they enhance the expressive power of object-oriented databases in two important directions.

Interfaces allow an object to be viewed in different ways by different users (human users or other objects). Triggers are mechanisms for causing side effects in other objects or at the user interface when an action occurs in a given object. In our system, a trigger can be associated with an arbitrary operation as opposed to LOOPS [BS83] in which triggers can be associated with *get-value* and *put-value* operations. They may be used to coordinate activities in a collection of objects, to synchronize state changes in strongly coupled objects that may represent the same physical component, and to alert users to actions that require a user response. Interfaces and triggers provide alternative mechanisms for defining multiple views of an object, as described in greater detail in section 5.

2.7. Control of the Level of Detail

Control over the level of detail of a specification can be illustrated by expanding the operations and attributes of a file to make the file more realistic and more UNIX-like.

```
Define Type File
  Operations:
    create-file () returns (F:File)
    open (F:File)
    close (F:File)
    rename (F:File, N:String) returns (F:File)
    copy (F:File) returns (F:File)
    set-protection (F:File, P:Protection) returns (F:File)
  Properties:
    device:Device
    file-name:String
    content (component): String
    owner-userid:String
    owner-group-id:String
    size:Integer
    date-last-read:Date
    date-last-written:Date
    date-last-modified:Date

Define Type Directory
Supertypes: File
  Operations:
    Create-file (D:Directory, N:Name) returns (D:Directory)
    Delete-file (D:Directory, F:File) returns (D:Directory)
    Create-directory (D:Directory, N:Name) returns (D:Directory)
    Lookup-file-name (D:Directory, N:Name) returns (F:File)
```

The above examples serve both to introduce the principal features of the object-oriented database language and to motivate their use. In the remaining sections we examine these features in greater depth and develop the beginnings of a methodology for using them to specify system objects of a program development environment.

3. ADA PACKAGES

In an execution time environment, Ada packages [SF] are simply objects to be executed. In a program development environment there are many additional things we may wish to do to Ada packages [GOGU83], such as edit and compile them. Moreover, we may wish to get at their components, for example to determine all packages that use a given library program, or all packages which have a private data type. The example below reflects this diversity of use. It specifies a package to be a type that can be created, edited, and compiled. Its properties include its name, the library units imported by with and use statements, the resources it provides to users, its private data types, and the package body component.

```
Define Type Package
   Operations:
      Create-Package (Name:String) returns (P:Package)
      Edit-Package (P:Package) returns  (P:Package)
      Compile (P:Package) returns (LM:LoadModule)
   Properties:
      Name: Strings
      With: Set of Library-units
      Uses: Set of Library-units
      Provides: Set of Resources
      Private:  Set of Declarative-items
      Package-body (component): Strings

Define Class Resources
Definition:
   Union [Subprograms, Packages, Variables, Exceptions, Types]

Define Class Library-units
Definition:
   Union [Subprograms, Packages]
```

This example includes some language features not previously explained. It illustrates that permitted values of a property can be defined in terms of a class or a set of class elements as well as by a type. It introduces ways of defining new classes as unions of classes. The keyword "component" indicates a property that stands in a particular relation to the type in which it occurs, namely the relation of being a component (the "has" relation).

The Ada package example assumes that the database contains object types for the program structures "subprogram, variable, exception" as well as for the system types "LoadModule, Type, String". It would not be difficult to write object-type specifications for the high-level program structures of Ada to create an object-oriented environment in which Ada programs could be edited using a system-supplied edit package and computed and executed using a system-supplied compiler. However, instead of discussing Ada in greater detail we will switch to Pascal and examine the object-oriented representation of program structure for this simpler language.

4. REPRESENTING PROGRAMS AS STRUCTURED OBJECTS

Programs may be represented in an object-oriented database as objects of large granularity at the level of Ada packages or complete programs or as objects of small granularity at the level of constants, variables, expressions, and statements. In this section we start by representing programs as objects of large granularity whose mechanisms for manipulation are hidden in elaborate system programs such as editors and compilers, and gradually introduce objects of smaller modularity that transfer more knowledge about program structure to the database itself and allow programs to be manipulated by database operations.

Object-oriented databases allow the level of detail of program representation in the database to be carefully controlled [ZW85]. The type *program* below represents programs as database objects that can be created, compiled and invoked, but relegates knowledge of internal program structure to the compiler and interpreter. This requires that the representation for this type be in a form that is recognizable by the compiler, but does not require that we make this representation available to users of this type.

```
Define Type Program
Supertypes: Procedure
  Operations:
    Create-Program (S:String) returns (P:Program)
    Invoke-Program (P:Program, Parameter-List:List)
    Compile(P: Program) returns (O:Object-Code)
  Properties:
    Compiles-into: Object-Code
    Load-Module: Executable
    Author: String
```

The operations *compile* and *invoke* are database objects whose implementation is a very simple use of the abstractions provided by the compiler and the operating system. The type *List* is a system defined type that is a subtype of the type *Ordered-aggregate*. It has all of the semantics that one would expect for lists. *Object-code* and *Executable* are types about which the database knows very little and may be defined by the following:

```
Define Type Object-Code
Inherits from: Program
  Operations:
    Link-and-Load (O:Object-code) returns (E:Executable)

Define Type Executable
Inherits from: Program
  Operations:
    Execute (E:Executable)
```

The operations, *Link-and-Load* and *Execute*, that are defined on these types are simple hooks to the facilities of the linker, the loader, and the operating system. Notice that our definition for *Program* includes several properties to relate the abstract program object to its associated object-code object and executable object. These properties would be set as a side-effect of the *invoke-program* operation which would in turn invoke the *Link-and-Load* and *Execute* operations.

The inherited type *procedure* may either be defined trivially as a type whose only role is inheritance of system properties, or substantively to reveal part of the internal structure of procedures. The following definition of the type Procedure indicates that programs, as well as procedures themselves, have environment and body components as well as a number of other properties.

```
Define Type Procedure
Supertypes: Entity
Associated Class: Procedures
  Operations:
    Create-Procedure (S:String) returns (P:Procedure)
    Invoke-Procedure (P:Procedure, A:List) returns (E:Entity)
    Display (P:Procedure)
  Properties:
    Name: Strings
    Environment (component): Set of Environment-Objects
    Body (component): Blocks
    Specification: Strings
    Documentation: Documents
    Last-modified: Dates
```

This definition has required us to factor out information that must be known at the program level from information that is known at the level of procedures. Operations such as compile that are not applicable to procedures must clearly be specified at the level of programs. But operations such as invoke could be omitted at the level of programs if there is a guarantee that they are specified at the level of procedures.

This definition of procedure types requires the database to know something about blocks. This type could be defined trivially by providing a type with no operations or properties for blocks. However, if we want the database to view a block as a collection of statement components, and to allow addition and deletion of statements from a block, then the type Block could be defined as follows:

```
Define Type Block
Inherits from: Statement
   Operations:
      Create-Block () returns (Block)
      Add-Statement (B1:Block, Pos:Integer) returns (B2:Block)
      Delete-Statement (B1:Block, Pos:Integer) returns (B2:Block)
   Properties:
      Statements (component): Set of Statment
```

In the above definition blocks both contain statements as components and inherit the type statement as a supertype. That is, they both "have" statements and "are" statements. This illustrates clearly the role of supertypes in realizing the "is-a" relation and the role of components in realizing the "has" relation. It also illustrates the fact that a given object may simultaneously have and be an object of a given type.

The type statements may be defined in terms of operations for execution and display and the property of having comments. Note that execution and display are thereby automatically inherited by particular statements such as blocks.

```
Define Type Statement (abbrev. Stmt)
Supertypes: Entity
Associated Class: Statements
   Operations:
      Execute (S:Stmt, E:Environment)
      Display (S:Stmt)
   Properties:
      Comments: Set of String
```

There are many different subtypes of statements in languages like Pascal. We have already encountered one subtype, namely the block. A further statement subtype, the conditional statement, is given below:

```
Define Type Conditional-Stmt (abbrev. CS)
Supertypes: Statement
Associated Class: Conditional-Stmts
   Operations:
      Create-CS (S:String) returns CS
   Properties:
      Condition (component):Predicate
      Then-clause (component): Statement
      Else-clause (component): Statement
```

Each object type definition provides more information about program structure at the database level. This information may be used by statistical programs, for example to determine how many conditional statements with an else clause there are in the program. A more complete definition of programs as hierarchical collections of objects could be used as a basis for writing the compiler in the object-oriented database language rather than in a traditional manner as a compiler whose internal structure is opaque to the database. The construction of compilers which view programs as hierarchical collections of database objects is being investigated.

4.1. Representing Programs in an Object-Oriented Database

An object-oriented database containing a set of program objects can be created in several ways. A user could explicitly invoke the *create* operations on the appropriate types to obtain the desired program object instances. These objects could in turn be used with other *create* operations to build up higher-level objects. For example, we could create a set of procedure objects that could then be used to build a main program.

Alternatively, one could build a language processor that takes textual descriptions of programs as input, parses this text into an abstract syntax tree, and then calls the appropriate *create* operations on behalf of the user. The output of such a compiler would be a set of database objects rather than object code. Object creation would take place bottom-up on the syntax tree.

Let us consider a simple example of how a procedure object would be created and represented. For this purpose, we will use the following procedure:

```
Procedure plus3 (VAR x:int);
  BEGIN
    x := x+3;
  END
```

The type definitions given above would require three objects to represent this procedure. One object would be created for the procedure itself, one object would be created for the local environment, and one object would be created for the body of the procedure. The creation of the local environment and the body would both require some representation of their content as input arguments. The most likely choice here would be strings. In this case, the local environment would be represented as "VAR x:int" and the body would be everything through and including the BEGIN END.

If this approach is taken, we are be left with a very simple view of programs and procedures. The internal structure of procedures is not available to the database system. For example, the database is unable to understand queries that talk about statements or variables. It would, however, be possible to share these objects in concurrent operating environments by means of the transaction facilities of the database system. It would also be possible to add information such as documentation to procedures.

Suppose that we want to articulate the semantics of programs to a much greater level of detail, and suppose that assignment statements, variables, and constants are all fully defined database types. This situation is depicted in Figure 1.

In this case, we would create many more objects for our above example. We would still create the three objects mentioned above, but in addition, we would also create an object of type *variable* for the environment and an object of type *assignment statement* as a part of the body. This assignment-statement object would be made up of two additional objects, the variable object "x" (the same object from the local environment) and an expression object that corresponds to the Pascal expression "x+3". The expression object would in turn be made up of two objects, the variable x and the constant 3.

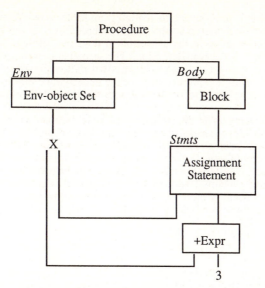

Figure 1 - Fully articulated program structure

With this more articulated structure, we are able to formulate more detailed retrieval requests. For example, the database can deal with requests like *Find all procedures that contain assignment statements that assign values to a variable named x.* It is possible to encapsulate specialized knowledge about program pieces with their corresponding types. For example, in graphical programming, different program pieces are displayed differently. By describing each program piece as a separate type, we can encapsulate the display operations with their corresponding type. This makes it very easy to modify the style in which things are displayed.

5. MULTIPLE VIEWS

Multiple views allow collections of objects to be accessed in different ways by different users. They may be implemented by objects which have multiple interfaces that share an underlying object representation. Changes via one interface must be instantly reflected in other interfaces. Interfaces are not fully fledged objects since their local data is shared by other interfaces to the same object, but should provide the user with the illusion that they are objects most of the time.

We examine two approaches to realizing multiple views in an object-oriented system. The first allows interfaces for a type to be defined at the time of the type's definition. It is conservative in the sense that it does not allow new interfaces to be defined independently of the type definition. The second approach allows each interface to be supported by a separate type with its own data representation and enforces synchronization of multiple representations of the underlying entity by "triggers" which allow lock-step updating of the set of all objects representing views of the same entity.

Interfaces and triggers are illustrated by multiple views of a program as a program text and as a Nassi-Schneiderman diagram. This type of problem arises in graphical programming systems such as PECAN [REIS84] or GARDEN [REISS] in which being able to manipulate any of several visualizations of a program is a central design goal. We would like the database system to assist us in keeping these two views synchronized. By describing the desired synchronization in the language of the database, this semantics will be available to other users of these types. The database allows us to share the semantics of the view interrelationships as well as individual objects.

5.1. Interfaces

Interfaces support multiple program views by a single type. The type *Programs* is defined to support two different interfaces. An interface is a subset of the operations and properties that are defined for the type as a whole. The interfaces give us a way to package and name interesting subsets that can be exported and used by other modules. Since the two interfaces are defined on top of a single type, they will both have access to the same underlying representation (i.e., the representation for objects of type *Program*). The operations of this type would be required to obey the representation invariant and thus transform the representation from one consistent state to another.

```
Define Type Program
Supertypes: Entity
Operations:
  Create-Program-Text ( ) returns (P:Program-Text)
  Display-Program-Text (P:Program-Text)
  Edit-Program-Text (P1:Program-Text) returns (P2:Program-Text)
  Execute-Program-Text (P:Program-Text, Args:List)
  Create-NS ( ) returns (NS)
  Execute-NS (N:NS, Args:List)
  Display-NS (N:NS)
  Add-New-Division (N:NS)
Properties:
  NS-Block (component): NS

Define Interface Program-Text (abbrev. PT)
Underlying Type: Program
  Operations: Create-Program-Text, Display-Program-Text,
    Edit-Program-Text, Execute-Program-Text
  Properties:

Define Interface Nassi-Schneiderman (abbrev. NS)
Underlying Type: Program
  Operations: Create-NS, Execute-NS,
    Display-NS, Add-New-Division
  Properties: NS-Block
```

The two interfaces are called *Program-Text* and *Nassi-Schneiderman*. Since there is a common object (i.e., the representation) that reflects changes made to either the Program-Text interface or the Nassi-Schneiderman interface, a change to one will necessarily produce a change to the other. As long as the representation invariant considers requirements imposed by both interfaces and all of the operations that modify either of the interfaces maintain this invariant, they will remain in synch. Of course, this does not necessarily insure that any displays of these two interfaces will be updated at the appropriate time. This is left up to the applications programmer.

Interfaces are used to export only the relevant semantics of a type to applications programs. A program that uses the database is always working within a *view*. A view is a subset of the types in the database (in this case, a single type) such that for each type T in the view, one of the possible interfaces T_i is selected. If a type has no alternative interfaces, the single type definition is selected by default. In the case in which there is only one type, a view is synonymous with an interface. In the more general case a view is constructed as a list of interface names for those types that are to be included in the view. A type definition is also considered to be an interface, thereby making it possible to list a type name for types which have no explicitly defined interfaces. There may be only one interface for a given underlying type in a given view. For the above, we might have:

```
Define View My-view
Definition: Program-text
```

An interface to a type T may select operations and properties from those defined on T. It may also define additional operations and properties in terms of functions on those that are available on T. For an *Employee* type with a *salary* property, we might define an interface that included a property called *FICA-payment* to be "0.05*salary".

It is tempting to consider supporting interfaces by a type hierarchy. With this approach, we would construct a common type T with one subtype for each interface that we wanted to support. T would have a representation that would be shared with all of the subtypes. Operations that are specific to some subtype T_i would be defined in the context of that type and would thus be available only to T_i. This appears to be the behavior that we want for interfaces, since T provides a common representation and operations are segregated to distinct program units via the subtypes. However, the operations on a subtype T_i cannot manipulate the shared representation of T directly, since only the operations defined directly on T are allowed to do that. This restriction is imposed by the need for proper modularity and data abstraction. Therefore, we must define all operations that need to access the representation of T as parts of T. A separate mechanism is needed to impose the notion of an interface on types. .

5.2. Triggers

Triggers define multiple views in terms of distinct types with no common representation. In this case, the correspondence between two objects which are, in fact, views of each other is expressed by means of explicit properties. The example shows the two types *Program-Text* and *Nassi-Schneiderman* with the properties *Corresponding-Nassi-Schneiderman* and *Corresponding-Program-Text* respectively. The value of each of these properties is an object that represents the alternative view

```
Define Type Program-Text (abbrev. PT)
Supertypes: String, Program
Associated Class: Program-Texts
  Operations:
   Create-Program-Text (T:String)
              returns (P:Program-Text)
   Create-Program-Text-from-NS (NS:Nassi-Schneiderman)
              returns (P:Program-Text)
   Display-Program-Text (PT:Program-Text)
   Edit-Program-Text (P1:Program-Text) returns (P2:Program-Text)
     Attach trigger Trigger-ProgText
   Execute (P:Program-Text, Args:List)
  Properties:
   Statements (component): Set of Statement
   Associated-Nassi-Schneiderman: Nassi-Schneiderman

Define Type Nassi-Schneiderman
Supertypes: Program
Associated Class: Nassi-Schneidermans
  Operations:
    Create-NS (LS:Set of Line-Segments) returns (N:NS)
    Create-NS-from-prog-text ( P:Program-Text) returns (N:NS)
    Display (N:NS)
    Add-New-Division (N:NS, B1:Block, B2:Block)
      Attach trigger Trigger-NS
    Execute (N:NS, Args:List)
  Properties:
    Block (component): NS
    Associated-Program-Text: Program-Tex
```

Since they do not share a common representation, changing one does not guarantee the appropriate change in the other. The types must define a *trigger* defined for each operation that modifies one of the views. The *trigger* type defines three fundamental properties: a predicate P, a block of code C, and a set of objects S with which the trigger is associated. A trigger T is *activated* whenever an operation is invoked on any one of the objects in the set S. When T is activated, if P is *true*, then C is executed. In the type Nassi-Schneiderman there is a trigger, Trigger-ProgText, associated with the operation *Add-new-division*. (the code for this trigger is defined elsewhere). Whenever this operation is invoked, the trigger is activated and the associated code is executed (In this case, P is simply equal to *true*). The trigger invokes operations on the other type (i.e., the program text) to perform the same modification on that type. Notice that the linkage between these two types is supported by the property called *Associated-program-text*, and that the trigger program would make use of this correspondence.

Notice also that there are two create operations for the two types. In both cases, there is a straight create operation and a "cross-type" create operation that creates an object of one type from an object of the other. A user first creates one of the two types using the straight create operation. The returned object is then used in the cross-type create operation for the second type. The cross-type create operation derives from its argument whatever information is required to construct an object representation for its own type.

6. CONCLUSIONS

We have examined the features that need to be added to an object-oriented programming language to produce an object-oriented database system capable of supporting a programming environment. Our approach is based on a data model and database system being developed at Brown. This system embodies a high-level semantic data model in the direction of [CHEN76, CODD79, ACO83, HM81, MBW80, SS77, ZDON84] that is strongly based on object-oriented programming concepts. We demonstrate a methodology for constructing such an environment by means of examples of object types that occur in the software engineering process. This approach provides a very general way of constructing integrated programming environments in which the database system provides the glue that holds together the set of tools and programming artifacts.

We have demonstrated that this approach has many strengths. The specification of the behavior of an object is localized to the module that defines its type. The level of abstraction that is achieved in the database can be very tightly controlled by specifying object types to the appropriate level of detail. System features can be easily included in the same way that we approach the constructing of applications. New relationships can be easily captured with this approach.

System design and application program development are both made easier with the use of an object-oriented database. Our system supports the incremental nature of these activities. Moreover, system design and application work can be approached using the same techniques.

Object-oriented programming methodology is not as well understood as programming methodology for traditional procedure-oriented languages. But object-oriented system and application programming may well be more effective than traditional styles of programming for large evolving systems and application programs. A great deal of work is needed both in implementing object-oriented database systems and in developing a methodology for using them. We hope that this paper contributes to persuading the reader that such an endeavor is worthwhile.

7. APPENDIX: RESEARCH AND OPEN QUESTIONS

In order to provide a broader perspective of our research in object-oriented database environments we list some of the research issues that need to be addressed in the realization of object-oriented database technology. The topics are taken from a more detailed description of projects and open questions made up for the benefit of master's and PhD students interested in further work in this area. This list of topics illustrates that the development of an object-oriented technology, just as the DoD-initiated effort to develop an Ada-based technology, requires a multi-faceted reseach program that cannot be accomplished by a single institution. One of the objectives of this paper is to suggest that object-oriented databse technology might provide as significant an increment in software productivity as was hoped for from Ada.

Programming Languages

- design of object-oriented databased programming languages.

- type issues in object-oriented programming.

- specification of object lifetimes.

- packaging system facilites so they can be easily inherited.

- comparison of object-oriented and traditional methodology.

- query language for object-oriented databases.

Databases

- consistency of object-oriented transactions.

- partial consistency for design transactions.

- transactions and version control.

- type specific concurrency control and recovery.

- triggers and data initiated actions.

- version control with changing types.

Environments

- design of an object-oriented programming environment.

- visual interface for object-oriented databases.

- integration of object-oriented systems into graphical environments.

- granularity of object-oriented specification.

- software tools for object-oriented programming.

- library structure for object-oriented programming.

- multi-paradigm languages and environments.

Implementation

- access methods and clustering techniques.

- embedded low-overhead objects.

- efficient storage strategies for version sets.

- making operations full fledged objects.

- garbage collection strategies for database systems.

- interactive, incremental debugging.

REFERENCES

[ACO83] [BS83] [BOUR78] [CARD85] [CHEN76] [CODD79] [CM84]
[CW85] [DH72] [GOGU83] [GR83] [HM81] [MBW80] [REIS84]
[REISS] [SF] [SS77] [ZDON84] [ZW85]

Chapter 12

PERSISTENCE AND ALDAT

T.H. Merrett
McGill University

ABSTRACT   *The Aldat project advocates the algebraic approach to data. This paper reviews this approach in the context of the persistent data type, relation, and the algebraic operators needed when including relations in a programming language. The relations discussed are the classical ones of Codd. The relational algebra consists of two families of binary operators and one of unary operators, and includes Codd's as a special case. We also review a domain algebra of two categories of operator. To illustrate the effectiveness of this extended relational algebra for programming, we show applications to text processing (concordance, document similarity coefficients), and geometry (point-in-polygon, intersection of lines). We introduce recursion into the extended relational algebra and discuss applications to logical deduction in Horn clauses and syllogisms and to the computation of property inheritance, such as arises in semantic typing in programming languages.*

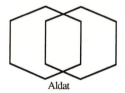
Aldat

The challenge of the next step in databases is either to say that relations are not the answer and move to some new model, or to extend the relational model with some (yet to be defined) new features.
                    Michael Stonebraker [1984]

1. INTRODUCTION

Aldat is a project to develop programming languages based on the algebraic approach to data. The main persistent data type we have investigated is the relation, which we have used to represent data in many diverse applications. The significance of the family of Aldat languages that has resulted is the algebraic operations on relations they embed, enabling the applications to be written at a very high level. These operations form an extension and generalization of Codd's relational algebra.   They allow administrative information systems, pictorial databases, text, and even knowledge bases to be implemented without any specific extensions to the language needed for any individual application.

The three important aspects of data on secondary storage are size, sharing and persistence. Database research has focussed largely on the first two, assuming the latter.

Operations on data which is too large to be brought into RAM have been the objective, first of query languages and, subsequently, of database programming languages. Since such large amounts of data are often a common resource, means were developed to permit concurrent access without rendering the data inconsistent. The fact that the data outlives the execution of any particular transaction or program was taken for granted until the work of the Data Curator project (see, for instance [ABCCM83]). This work introduced the principle that *all* data objects have the same rights to persistence or transience, and built languages (PS-Algol, Nepal) which support persistence of any data type. Our present purpose is to demonstrate just how far we can get with a sole data type, the *relation* of database research. The value of doing this in the present context is to complement the approach which uses many data types. Data types other than relations are needed for sophisticated work, and to them the principle that persistence is "orthogonal" - independent of data type - must certainly apply. We do not discuss these but content ourselves with exploring the applicability of relations alone. Their persistence poses no new implementation problems.

Persistence is hardly discussed explicitly in this paper, because of our database perspective: that relations persist is assumed in all implementations we have made of Aldat, and we supply only an explicit **delete** operator to remove any named relations which we do not want to keep. Large amounts of data cause no difficulty, since all operators introduced are database operators and intended for large databases on secondary storage. Sharing is also not explicitly discussed because that takes us into the province of editors, which we only briefly mention, and into aspects of concurrency control and absence thereof which are the concern of on-going work. This paper reviews the Aldat formalism and extensions to the relational algebra. Our main objective is to establish the advantages of Aldat programming on persistent data by developing examples from some of the areas of current interest in database programming. Administrative data processing having been extensively investigated in the past decade and a half, we look instead at the more recently interesting topics of text, pictures and logic programming.

Our interest in all cases is not just the representation of the appropriate data but the operations which permit us to process the data. In formulating operations we adhere to two *algebraic principles*: the principle of *closure* requires that operations on a type of object result in objects of the same type; and the principle of *abstraction* requires that neither the structure of objects nor their context should be of concern to the operations applied to them. Relations are very suitable as objects to represent persistent data under these algebraic principles, because they usually represent quantities of data at least the size of a physical block and because the principle of abstraction tends to allow us to process whole blocks at a time. This leads us to implementation considerations, which are important in our work but which we do not discuss further here.

In the first two sections, we summarize the extended relational algebra and the domain algebra respectively. Section 3 gives programming applications to text processing and to geometrical computations. For text, we compute the concordance of a document and the similarity coefficients of several documents, a first step in cluster analysis. For diagrams, we give a point-in-polygon algorithm and we calculate the intersection points of two closed lines. In section 4, recursive relations are introduced. Recursion is illustrated for classical fixed-point problems such as bill-of-materials and routing, then for logic programming to do both Horn clause and syllogistic deductions. Finally, the problem of property inheritance, as it arises in semantic networks and data types, is solved as a closure computation in the relation algebra.

2. OPERATIONS ON RELATIONS

The relational model is grounded in set theory - a *relation* is a set of tuples, or, more precisely, it is a subset of the cartesian product of its domains (the sets from which its attributes are drawn). A set is a special case of a relation, namely a unary relation. It is appropriate that operations on relations should generalize the operations already provided for sets. There are three classes of such operations: unary operations (complementation), binary operations which result in sets (union, intersection, etc.) and binary operations which result in logical or boolean values (inclusion, disjointness, etc.)

The unary operation of complementation does not seem to be very useful for relations, which are stored explicitly in a computer and which are usually sparse relative to the universe of all possible tuples. Complementation has not been used in queries or data processing because the notion of the universe is not free from ambiguity in the user's mind: we shall say more on this when we discuss quantification. It is the identity operation on sets which generalizes to the relational operations of project and restrict. (Sorting is such an identity operation, since the abstraction which created the notion of set ignores the order of the elements.) The binary set operations give easier generalizations, and we discuss them first.

2.1 Binary Operations

μ-**joins** Both categories of binary operation on sets - those which produce sets and those which produce booleans - extend to relational operators which produce relations. We start with the first, the operations of union, intersection, difference and symmetric difference, and look at the natural join [CODD70] in their light. If we specialize the natural join to operate on sets, we have exactly set intersection. We thus call natural join *intersection join*. There are conceivably many ways to generalize set union to relations, but it seems appropriate to generalize it in the same way that intersection extends to natural join. This has been done by a number of people, to give the *outer join* - see [DATE83] for some references. Define the *center* to be the natural join of relations $R(X, Y)$ and $S(Y, Z)$, here written $R \cap S$. Define the *left wing* to be those tuples of R which do not participate in $R \cap S$, augmented by null values for the attribute Z. Define the *right wing* similarly as the tuples of S which match no tuples in R, augmented by null values for attribute X. The *union join* (outer join) is just the set union of left wing, center and right wing. It specializes to set union. Similarly, we can extend the other set operations, and add the *left join*, which specializes to an identity operation.

Alternative Notations

```
R ∩ S = center    (natural join)                          R ijoin S
R ∪ S = left wing ∪center ∪right wing   (outer join)      R ujoin S
R + S = left wing ∪ right wing                            R sjoin S
R - S = left wing                                         R djoin S

              (here it is convenient to project the
              left wing on attributes X and Y)

R left S  = left wing ∪ center                            R ljoin S
```

This family of joins, generalizing the set-valued binary operations on sets, is called the μ-*join*.

σ-joins Relational division, presented by [CODD71a] as an algebraic counterpart to the universal quantifier, is a little hard for many people to understand. It becomes clearer when perceived as a generalization of set inclusion. For relations $R(X, Y)$ and $S(Y)$, the division $R \supseteq S$ is just those values of X that are associated by R with a set of values of Y which contain the set S, $R \supseteq S = \{x|\, R_X \supseteq S\}$ where R_X is the set of Y-values associated with x in R. This can be extended by allowing any set comparison, such as $\supset$, $\not\supseteq$, $=$, etc., to replace $\supseteq$. It can be further extended to relations $R(X, Y)$ and $S(Y, Z)$ by defining

$$R \supseteq S = \{(x,z)|\, R_X \supseteq S_Z\} \text{ with } S_Z \text{ having a meaning similar to } R_X.$$

For completeness, we must introduce two new set comparisons and their complements:⋒ tests the two sets compared for empty intersection and ⋓ tests whether they span the universe. The operation R ⋒ S is the *natural composition* [CODD70], a most useful operation in its own right. Although it is definable as a projection of the intersection join, it is hereby seen to be a closer cousin to division than to the natural join. An alternative notation, which we use for ⋒, is **icomp**

This family of joins is called the σ-*join*. If we allow a relation with no attributes to be a *scalar*, taking on the boolean values **true** or **false**, we see that the σ-joins specialize to the corresponding set comparison operators.

2.2. Unary Operations

QT-selectors The unary relational operators of *projection* and *selection* can be conveniently combined into a single operator with a simple syntax: for instance,

$$W, Y \text{ where } (X < \text{'tim' or } Y = 2) \text{ in } R$$

is a selection $(X < \text{'tim' or } Y = 2)$ on a relation $R(W,X,Y,...)$ followed by a projection on the attributes W and Y. The selection condition can be anything at all, provided it can be evaluated **true** or **false** on each tuple of the relation without reference to other tuples. It is called a *T-condition* (T for "tuple") for this reason, and the operation is a *T-selector*.

The *QT-selector* generalizes this to include *quantifiers* in the selection condition. The quantities envisaged are not limited to the classical "for some" and "for all". The two quantifier symbols #("the number of") and • ("the proportion of") are used in quantifier expressions to generalize "for some" and "for all" . Thus

$$W, Y \text{ where } (\# > 1) Z, (X < \text{'tim' or } Y = 2) \text{ in } R,$$

applied to $R\ (W,X,Y,Z,...)$, reads "find W and Y where, for at least one Z, $X <$ `tim' or $Y = 2$ in R"; and

$$X, Y \text{ where } (\bullet \leq .5) \ W, Z \neq X \text{ in } R$$

reads "find X and Y where, for no more than half of the values of W, $Z \neq X$ in R". The special cases "for some" and "for all" correspond to the quantifier expressions, (#>0) and (•=1), respectively. Quantifiers may be combined in one QT-selector, and obey the rule, followed by their set-theoretic special cases, that changing the order of the quantifiers generally changes the meaning of the QT-selector.

While the QT-selector looks like a general query language facility, and exceeds most query languages in functionality, it is only a unary operator on relations, satisfying the algebraic requirement of closure, namely that its result is in turn a relation. Thus QT-selectors may be nested or combined with other relational operators such as µ-joins and σ-joins.

A word about the universal quantifier is advisable here. When we say "for no more than half the values of W", do we mean relative to all values of W that appear in R, or something else? This is the problem of ambiguous - or just hard-to-calculate - universes. For various practical reasons, we define the universe in terms of the relation (or relational expression) appearing after in the **in** keyword.

QT-selectors can be used to identify subsets of a relation to be updated, using a more extensive notation. We look next, however, at a simple operator permitting tuple-at-a-time inspection and updating of a relation.

Relational Editor The operations of the relational algebra discussed so far have obeyed two fundamental algebraic principles. The first is the principle of *closure*, that a relational operation produce a relation as a result. This permits the construction of a relational expressions consisting of one or several operations. The second principle is *abstraction*, which permits us to ignore the internal structure of relations. Thus we have been able to ignore the tuples which constitute the relations we have been operating on. Our notation has not even required tuple variables, range statements or *for each* loops, which explicitly focus attention on the microstructure of relations in many systems and notations. For the person responsible for the design of an information system, freedom from this distraction is ideal. This is the person who applies the formalism to a particular problem. We will call him the "programmer user".

For the person who actually uses the data, however, the tuples are of primary interest and the fact that they are grouped into relations is only incidental. This person enters the data and reads the reports. We will call her the "end user". One important activity for the end user is to examine and possibly change the tuples using an interactive editor. How can we reconcile the two complementary views the programmer user and the end user have of the data in a single operation?

Our answer is a two-faced operator. To the programmer user, the relational editor looks like a simple unary relational operator, like project for instance. A simple syntax such as **edit** R will result in a new relation (which may be assigned back to R). Only instead of the operation being defined algorithmically, as would be the case for the projection, X,Y **in** R, the output relation is a result of the free activity of the end user at an interactive terminal using an appropriate command language. This interactive command language is the second face of the **edit** operator, the face presented to the end user, who can thereby see and edit the tuples of her data.

• • •

The foregoing is an outline of an extended relational algebra which is discussed precisely and more fully by [MERR84]. A system including all the above operations has been implemented in U.C.S.D. Pascal on an Apple II [CHIU82].

3. OPERATIONS ON ATTRIBUTES

Many practical problems for databases require computations to be performed on attributes. For instance, multiplying an income by a tax rate for every tuple in a salary relation, or totalling sales by department. Most implemented systems handle at most some of the possible requirements, and these in an *ad hoc* fashion, such as by special functions for COUNT, TOTAL, AVERAGE, etc. The failure of these approaches to adopt a framework for operations on attributes which is consistent with the operations they provide for relations leads to dislocations of syntax and to incompleteness.

To be consistent with the relational algebra, we must embody our operations on attributes in an algebraic framework and, in particular, we must observe the principles of closure and of abstraction. Closure requires that we operate on attributes to get attributes and abstraction requires that we avoid considering the structure of attributes or their connection with particular relations. The resulting formalism we call the *domain algebra*, and it comes in two flavours, horizontal and vertical.

3.1 Horizontal Operations

Horizontal operations are defined on any one tuple and are applied repeatedly to each tuple in a relation. They can be any arithmetical, logical or other operation or expression. They are "horizontal" because they can be imagined to operate horizontally along each row of the usual tabular representation of relations. In keeping with our algebraic principles, of course, there is no mention of tuples or of relations in the syntax:

```
let MARK be MIDTERM + FINAL + ASSIGNMENTS
```

creates an attribute MARK from the existing attributes MIDTERM, FINAL and ASSIGNMENTS.

Since MARK has been defined above in the absence of any particular data, it is a *virtual* attribute until it is *actualized* in the context of some particular relation. The most obvious mechanism for actualization is an operation of the relational algebra, such as projection. Thus, if we have a relation *COURSE (STUDENT, MIDTERM, FINAL, ASSIGNMENTS)* we could project the relation:

```
STUDENT, MARK in COURSE.
```

This syntax can also allow us to define constant attributes, as in

```
let ONE be 1
```

and to rename attributes, as in

```
let newname be oldname.
```

3.2 Vertical Operations

If horizontal operations work "along the tuples", vertical operations work "down the attributes", and fill the roles of totalling, subtotalling, integrating and partial integrating and their generalizations. For the first we have *reduction*:

```
let TOTAL be red + of MARK
let COUNT be red + of 1
let AVMK be TOTAL/COUNT          << a horizontal operation >>
let MAXMK be red max of MARK
```

for the second we have *equivalence reduction*:

```
let SUBTOT be equiv + of MARK by SECTION
let SUBCT be equiv + of 1 by SECTION
let SUBAVMK be SUBTOT/SUBCT      << a horizontal operation >>
let SUBMXMK be equiv max of MARK by SECTION
```

for the third and fourth we have *functional mapping* and *partial functional mapping*:

```
let INTF be fcn + of F order X
let PINTF be par + of F order X by Y
```

We do not discuss these latter except to remark that, in the above simple forms, they provide poor integrals, but can be improved to perform as well as the data will allow. Note the ordering clause, which effectively specifies a sort on a given attribute or set of attributes, and makes very useful the special operations **pred** (predecessor in the ordering) and **succ** (successor in the ordering):

```
let NEXTEVENT be fcn succ of EVENT order TIME.
```

While these vertical operations of the domain algebra were motivated by practical considerations arising from the application areas we have investigated in our research, fundamental principles and basic mathematical concepts have been used in their definition. Thus, apart from the principles of algebraic closure and algebraic abstraction, we have applied the notion of "equivalence class" in equivalence reduction and the notion of "functional" in functional mapping. Note that the **order** clause in functional and partial functional mapping does not violate the relational abstraction that order does not matter. Functional mapping cannot, for instance, be used to print out a given relation in order: that must be done by a specialized print routine, which is not of particular interest to the relational algebra because it is at best an identity operation.

• • •

The foregoing is an outline of a domain algebra which is discussed precisely and more fully by [MERR84]. A system including this domain algebra and the relational algebra of the previous section has been implemented in U.C.S.D. Pascal on an IBM Personal Computer [VANR83], [DHIL85].

4. PRACTICAL JUSTIFICATION

While the above discussion has been couched in terms of principle and mathematical concept, each extension and generalization we have made to Codd's original relational algebra has resulted from a practical inadequacy of an existing system. We have investigated, in generalized and somewhat abstract form, commercial, library, geographical and text information systems. Commercial systems discussed in [MERR84] include manufacturing and financial information systems. We have investigated both administration systems for libraries - *e.g.*, acquisition and circulation - and information retrieval aspects of document clustering and search. Geographical databases have given us a framework for the study of information systems based on large two-dimensional images and diagrams. Text processing has been examined in a variety of ways, including editing and page formatting only as two of a broad range of processing: indexing, transliteration, linguistic analysis, encryption, etc. The formalisms of the relational and domain algebras have been adequate to handle all the well-defined operations of these widely differing semantic contexts, although they strain a little around the intricacies of document classification and search, and special routines must be provided, resembling the relational editor, for the interactive operations of picture and text editing.

4.1 Text Processing

A very good reason for representing text as a relation is the variety of computations we can perform on it using the relational algebra. Text processing should not be limited to editing and formatting (page layout, typesetting). To represent text - which can in its simplest essence be seen as a sequence of words - in relational form, we must add a sequence number. Any sequence can be represented as a set by extending the set elements to include sequence numbers, despite the apparent differences between sequences and sets (order does not matter in a set; duplicates are allowed in a sequence). An encouraging consequence of this approach is that a *text is its own concordance*, if we take a concordance to be an index to all words in the text: as a text, the data is usually arranged in sequence number order; as a concordance it appears in alphabetical order of words. Here is a text, a list of "stop words" - insignificant but frequently occurring words - and an improved concordance, calculated by removing stop words using the difference join.

CONCORDANCE <- CORPUS **djoin** STOP

CORPUS (WORD	DOC	SEQ)	STOP (WORD)	CONCORDANCE (WORD	DOC	SEQ)
Little	A	1	a	baa	B	1
Bo	A	2	alone	baa	B	2
Peep	A	3	and	bags	B	14
has	A	4	any	black	B	3
:	:	:	behind	bo	A	2
Baa	B	1	come	bowl	D	25
baa	B	2	does	boy	B	28
black	B	3	for	called	D	17
sheep	B	4	has	called	D	21
have	B	5	have	called	D	28
:	:	:	he	cat	C	5
			:	:	:	:

We can begin a process of clustering the different documents in *CORPUS* by calculating *term vectors* - a list of words and their relative frequencies - and a *similarity matrix* associating the documents. The term vectors hold normalized frequencies and the similarity coefficient used is the cosine coefficient or scalar product of the normalized frequencies. Note that we must find the natural join of *TERMVECT* with itself on *WORD* in order to calculate the contribution of each word to the cosine coefficient. The equivalence reduction used to find *COSINE* sums these contributions over all words common to each pair of documents. The assignment creating i just copies *TERMVECT*, renaming attributes so the join can be done correctly.

```
let WORDFREQ be equiv + of WORD, DOC
let NORMFREQ be WORDFREQ/sqrt (red + of WORDFREQ**2)
TERMVECT <- WORD, DOC, NORMFREQ in CONCORDANCE
```

```
TERMVECT (      WORD        DOC        NORMFREQ )
                baa         B          2
                bags        B          1
                black       B          1
                Bo          A          1
```

```
let COSINE be equiv + of NORMFREQ*NF1 by DOC, DOC1
TERMVECT1 [WORD, NF1, DOC1
          <- WORD, NORMFREQ, DOC] TERMVECT
DOCSIM <- DOC, DOC1, COSINE where DOC ≠ DOC1
          in (TERMVECT ijoin TERMVECT)
```

Here is *DOCSIM* in matrix form. Blank entries are zero and the matrix is symmetric.

		DOC1		
DOC	A	B	C	D
A		.13	.07	
B	.13		.05	.04
C	.07	.05		
D		.04		

Other text applications are described by [MERR85a].

4.2 Geometrical Computations

A *diagram* is a line drawing, as used in maps, charts, graphs and technical illustrations. It is two dimensional and contains *points*, *lines*, and *regions* as elements. We will limit our attention to *polygonal* representations of diagrams, in which lines and the boundaries of regions are approximated by straight line segments. This representation has the advantages that polygons have been the recent subject of intense investigation in computational geometry and that they can be stored as *sequences* of points, so that all three types of element (point, line and region) can be represented uniformly. (However, splines can also be used to fit sequences of points, as in Metafont [KNUT79], so we can represent curves, but we do not discuss this further here.)

A *feature* is a diagram or a subdiagram identified by a name, consisting of one or more points, lines or regions, i.e., *groups* of sequences. A relational representation of the Aldat logo, at the beginning of this paper, is shown in *DIAGRAM*. The feature 'logo' is considered in this example to be a pair of regions: the assumption of the **pred** and **succ** operators defined in Section 2.2 is that ordering is cyclic, so that closed lines are easy to achieve. To leave a line open, the tuple connecting the highest and lowest sequence numbers of each group must be found and removed after any **succ** or **pred** operation. A region is always bounded by a closed line. A point is a singleton sequence. Note that the representation in *DIAGRAM* does not cope with the hierarchy of picture, sub-picture, sub-sub-picture, etc., which is important in many diagrams, but this is an easy extension. Note also that *DIAGRAM* is not in third normal form [CODD71a]: the presence or absence of this normalization is not of concern here.

DIAGRAM(	FEATURE	TYPE	GROUP	SEQ	X	Y)
	logo	region	1	1	-.5	-1
	logo	region	1	2	.366	-.5
	logo	region	1	3	.366	.5
	logo	region	1	4	-.5	1
	logo	region	1	5	-1.37	.5
	logo	region	1	6	-1.37	-.5
	logo	region	2	1	.5	-1
	logo	region	2	2	-.366	-.5
	logo	region	2	3	-.366	.5
	logo	region	2	4	.5	1
	logo	region	2	5	1.37	.5
	logo	region	2	6	1.37	-.5

Some geometrical queries which could be made on *DIAGRAM* are the following. We show the algebraic operations to answer some of them. "Point" means any point already in the database. "Arbitrary point" means any point. Note that we assume, given a line, 'name', or a region, `name`, that this is the only line or region in the feature 'name'. Questions 7 and 8 raise some technicalities from computational geometry which we do not elaborate on here.

1. Find a given feature, 'name' (point, line or region)
   TYPE, GROUP, SEQ, X, Y **where** FEATURE = 'name' **in** DIAGRAM

2a. Given a point, (PX, PY), find what line it is in.
   FEATURE **where** (TYPE, X, Y) = ('line', PX, PY) **in** DIAGRAM

2b. Given a line, `name`, find what points are in it.
   X, Y **where** (FEATURE, TYPE) = ('name', 'line') **in** DIAGRAM

2c. Given a region, `name`, find its boundary.
   SEQ, X, Y **where** (FEATURE, TYPE) = ('name', 'region') **in** DIAGRAM

2d. Given a closed line, find the region it bounds.
3a. Given an arbitrary point, (PX, PY), find what region it is in.
3b. Given a line, find what region it is in.
3c. Given a region, find what points are in it.
3d. Given a region, find what lines are in it.
4a. Given two lines, `name 1' and `name 2', find their intersection.
4b. Given two regions, find their intersection.
5a. Given a line and two arbitrary points on it, find the length of the line between the points.
5b. Given a closed line bounding a region, find the area and centroid of the region.
6a. Given arbitrary point, find the nearest point.
6b. Given an arbitrary point, find the nearest line.
6c. Given a line, find the nearest different line.
6d. Given a line, find the nearest point not on it.
7. Given a region, find its convex hull, visibility graph, medial axes, triangulation...
8. Given a set of points, find the Voronoi diagram and the Delauney triangulation.

We elaborate briefly on queries 3a and 4a. We use the sum of angles method for 3a, which says that angles of rays drawn from point to all vertices of a polygon sum to $\pm 2\pi$ if the polygon contains the point, and to 0 otherwise. The angle is calculated using $C^2 = A^2 + B^2 - 2AB \cos \emptyset$ where C is the length of the subtended side of the polygon and A and B are the lengths of the adjacent rays. This does not give the sign of $\emptyset$, the angle, which is positive or negative depending on whether the order of the points (X,Y), $(X´, Y´)$ and

(PX, PY) is counterclockwise or clockwise, where *(X´, Y´)* is the cyclic successor of *(X,Y)*. This is given by the sign of the determinant of the 3 * 3 matrix

```
1    X    Y
1    X´   Y´
1    PX   PY.
```

```
let A2 be (X - PX)**2 + (Y - PY)**2
let X´ be par succ of X order SEQ by GROUP
let Y´ be par succ of Y order SEQ by GROUP
let B2 be par succ of A2 order SEQ by GROUP
let C2 be (X-X´)**2 + (Y-Y´)**2
let ANGLE be arccos ((A2 + B2 - C2)/(2 * sqrt(A2 * B2)))
let AREA be det3(1,X,Y,1,X´,Y´,1,PX,PY)
let TOTANG be equiv + of sign(AREA) * ANGLE
        by FEATURE, GROUP
FEATURE, GROUP where abs(TOTANG) = 2 * π and
        TYPE = 'region' in DIAGRAM
```

A test with point (-.5,0) against the Aldat logo shows that the point is contained in the first group (hexagon) but not the second. Point (0,0) is in both.

In query 4a, we must combine every edge of the first line with every edge of the second line. This involves selecting the lines and finding their cartesian product. Note how we use **ijoin** for this. Edges, of course, are determined by pairs of points generated by the successor operation. Once all combinations are found, we can compute the intersection coordinates for each combination and determine whether they lie within the base defined by the four endpoints of the two intersecting edges. We assign the result coordinates the "don't care" null values, DC [MERR84], if the lines are parallel.

```
let X´ be par succ of X order SEQ by GROUP
let Y´ be par succ of Y order SEQ by GROUP
let XX be X; YY be Y;  XX´ be  X´; YY´ be Y´
let A be X-X´; B be Y´ - Y; C be Y *A + X*B
let AA be A; BB be B; CC be C
let DEN be B*AA - BB*A
let YP be if DEN = 0 then DC else  (B*CC - BB*C)/DEN
let XP be if DEN = 0 then DC else if B = 0 then (CC - AA*YP)/BB
        else (C - A*YP)/B
XP,YP where X min X´≤ XP ≤ X max X´ and
            XX minXX´ ≤XP ≤ XX max XX´ and
            Y min Y´ ≤ YP ≤Y max Y´and
            YY min YY´ ≤ YP ≤ YY max YY´in (
(X,X´,Y,Y´,A,B,C where (FEATURE, TYPE )= ('name 1','line')
        in DIAGRAM) ijoin
(XX,XX´,YY,YY´,A,BB,CC where (FEATURE, TYPE) = ('name 2','line')
        in DIAGRAM)
```

This code works for closed lines, such as the hexagons in 'logo'; they intersect at points (0,-.711) and (0,.711). For open lines, we must add to the selection conditions before the **ijoin** the proviso that SEQ not exceed its successor.

While pursuing the polygonal approach to geometrical data, we have not mentioned its major rival, the grid or pixel approach. A rectilinear grid of picture elements ("pixels"), each with a variable grey level (or colour level), is very suitable for wirephotos, television images, landsat data, etc. It is less flexible for processing but has the advantage of being able to localize interesting parts of the picture very well because of the simple cartesian coordinate system it provides. Our research into storage structures for relations has produced multipaging [MERR84], which permits a hybrid of polygon and grid data structures for diagrams, with the best features of both representations [DUEC83]. A survey of literature on relational representation of pictures is given by [MD85].

5. RECURSIVE RELATIONS

The recursive definition of ancestor in Prolog is well known. Here it is in Aldat:

```
relation ancestor, parent (senior:person, junior:person);
<< declare ancestor, parent, as relations on (senior,junior) >>
ancestor <- parent ujoin parent [junior icomp senior] ancestor.
```

This expression requires no new syntax but introduces the meta-feature that recursive definitions of relations are allowed. Simple implementation is possible in this case. The above definition is equivalent to:

```
ancestor <- { };                              <<empty relation>>
repeat test <- ancestor;
ancestor <- parent ujoin parent [junior icomp senior] ancestor
until (test = ancestor);
```

It is helpful that the central statement, embedded in the loop, is identical to the recursive definition implemented by the above code. This makes it easy to generate the code mechanically from the easily-understood concept that an ancestor is a parent or the parent of an ancestor.

In what follows, any recursive definition or set of mutually recursive definitions will be assumed to be implemented by embedding them in a loop with the two characterstics satisfied by the implementation, above, of ancestor: relations are initially empty, and the loop is executed until there is no further change in any of the recursive relations. This permits any arbitrary recursive definition of a relation, as long as the recursion implies monotone growth of the relation while the loop is being executed. (This monotonicity is guaranteed for the classical relational algebra of project, select, union and natural join. It is not for full recursion in a formalism which includes, for instance, **djoin**. We have an implementation which allows non-empty initialization, but do not discuss it here.) The implementation is, of course, naive (see [BANC85]), but we are concerned here with generality of definition rather than efficiency of implementation.

The above recursive definition applies to many different situations. For instance, a bill-of-materials or part-of hierarchy frequently needs to be ``exploded'' into all its component parts. We could have:

```
relation BOM, components (assembly:part, subassembly:part);
components <- BOM ujoin BOM [subassembly icomp assembly] components.
```

Similar observation can be made about air routes connecting city to city and the set of all possible paths that can be created with zero or more stopovers:

```
relation route, path (from:city, to:city);
path <- route ujoin route [to icomp from] path.
```

And so on. These are all examples of finding the *transitive closure* of a graph, which in turn is a classic *least fixed-point* problem.

The bill-of-materials example is interesting because some arithmetic computations may be performed along with the topological computation of the transitive closure. For instance, if we associate with each assembly-subassembly connection the *quantity* of that particular subassembly needed in constructing its parent assembly, we can compute overall quantities in the explosion. We notice that we need multiplication of quantities along paths, and addition of quantities where parallel paths with common endpoints meet. For example, if A consists of 2 Bs and 3 Cs, and B and C each consist of 5 Ds, then A consists of 2*5+3*5 = 25Ds. Here is the computation:

```
relation BOM, components (assembly: part, subassembly:part,
        qty:+ve integer);
let a be assembly; s be subassembly; q be qty;
let qq be equiv + of q * qty by (assembly, s); qqq be
                                              q + qq;

components <- assembly, subassembly, qqq in BOM
        [assembly, subassembly ujoin assembly, s]
        (assembly, s, qq in (a,s,q in BOM)
        [a ijoin subassembly] components)
```

Some working of examples is needed to appreciate this code, but the computation is equivalent to a page of Prolog [CM81, p. 120]. Furthermore, given components, the total cost of assembling any of the components can easily be found if we have a relation telling us the assembly cost for the component and for its subassemblies.

5.1 Horn Clauses

Recursive relations are thus valuable in quite common information processing applications. We return, however, to logic programming. A class of logical deductions which can be made using recursive relational algebra expressions involves Horn clauses. A Horn clause is a rule with several antecedents, to be **and**ed together, and one conclusion. Representing a set of Horn clauses as a ternary relation is easy. If we also have a set of facts, we can apply the Horn clauses by selecting rules all of whose antecedents are contained in the set of facts, and then add the conclusion to the known facts. This can be done repeatedly, or recursively, as follows:

```
relation Horn (rule:name, ante:predicate, concl:predicate);
        facts, newfacts (concl:predicate);
newfacts <- facts ujoin concl in (newfacts [concl ⊇ ante] Horn).
```

Here the σ-join ⊇ is used instead of **icomp** to ensure that *all* antecedents of a given rule are found in the known facts before its conclusion can be added to the facts. The iterative implementation of the recursive definition of newfacts follows the same pattern as that shown above for ancestor.

We illustrate this representation and computation with an example.

Horn(rule	ante	concl)	facts(concl)	newfacts(concl)
1	lays eggs	is bird	lays eggs	lays eggs
1	has feathers	is bird	has feathers	has feathers
2	flies	is bird	swims	swims
2	is not mammal	is bird	is brown	is brown
3	is bird	is duck		is bird
3	swims	is duck		is duck
3	is brown	is duck		

Notice the difference between the expression for newfacts above, which can be thought of as the closure of facts under the Horn rules, and that for *ISA*. We have not computed the closure of the rules, say *Horn**, as we do *ISA**, below. Because a set of Horn clauses involves both **and** and **or** operations on its antecedents (see rules 1 and 2 above), the number of rules which can be derived from a given set may grow exponentially in the length of the derivation. (Try combining rules 1 and 3, rules 2 and 3 above to characterize ducks directly.)

5.2 Syllogisms

The closure of an *ISA* relationship [SS77] can be computed using code identical to that for ancestor:

```
relation ISA, ISA* (subject:class, object:class);
ISA* <- ISA ujoin ISA [object icomp subject] ISA*.
```

The *ISA* relationship is just the universal part of the syllogisms of classical logic, and has two rules for computing conclusions from a given set of *ISA* relationships. One rule is *transitivity*, hence the above closure. The second rule is *antisymmetry*, and applies to convert an *ISA* relationship on classes to an *ISA* relationship on the *complements* of those classes. Thus:

```
dog ISA mammal ≡ not mammal ISA not dog
```

To include this rule in the closure, we can find first the closure of ISA under antisymmetry then the closure of this under transitivity, since it can be shown that the transitive closure of an antisymmetric ISA is itself antisymmetric. (We cannot use the faster computation of transitive closure followed by antisymmetric closure because it does not always work, as in *a ISA b, c ISA d, b ISA* **not** *d*.) The relations must change slightly:

```
relation ISA, ISA* (subj:class, sc:boolean, obj:class, oc:boolean);
let sc´ be not sc; let oc´ be not oc;
ISA[subj, sc, obj, oc <+ obj, oc ´, subj, sc ´] ISA
     << incremental assignment with renaming >>
ISA* <- ISA ujoin ISA [obj, oc icomp subj, sc] ISA*
```

The above extends the ISA heirarchy to the universally quantified syllogisms and can be used to deduce any of the classical forms Barbara, Celarent, Cesare or Camestres. The other syllogistic forms include existential qualification, as in "some dogs are brow'". We call this relation *LAPS*, for "overlaps", as in "dogs *LAPS* brown things": the set of dogs overlaps the set of brown things. There are four rules, three of which relate *LAPS* to *ISA*:

symmetry: *a LAPS b <->b LAPS a*
specialization: *a ISA b -> a LAPS b*
intersection: *a ISA b* **and** *a ISA c -> b LAPS c*
composition: *a LAPS b* **and** *b ISA c -> a LAPS c*

It is easy to write relational algebra expressions for each of these, and to use these to compute, if we wish, all syllogistic conclusions from a given set of ISA and LAPS relationships.

4.3 Inheritance

The notions of classification (MEMBER_OF), generalization (ISA) and aggregation (HASA) are of growing interest to researchers linking databases, programming languages and artificial intelligence. Inheritance of properties (HASA) within a generalization hierarchy (ISA) is an important aspect. We see this in [BMS84] in the papers by [BMW84], [REIT84] and [BR84]; and inheritance in the specific context of programming language types is discussed in the present workshop by [AM85], [AGOP85], [ZW85], [BUNE85] and [AITK85].

Inheritance can be computed by finding the closure of HASA under repeated application of ISA in the following way. We find the reflexive-transitive closure of *ISA*, *ISA**, using the reflexive part, *ISA* °

```
relation ISA, ISA°, ISA* (subject:class, object:class);
HASA, HASA* (object:class, quantity:property);
let obj be object; let subj be subject;
ISA °[object,subject <- object, obj] (object,obj in ISA)
          ujoin (subject, subj in ISA)
ISA* <- ISA° ujoin ISA [object icomp subject] ISA*
```

Then we compute the inheritance of all properties in *HASA*:

```
HASA* [object, quality <- subject, quality] ISA* icomp HASA
```

Note that multiple inheritance is handled as easily as single inheritance in this computation. Exceptions, such as flightless birds, must be noted explicitly, say in a relation *HASNOT (object, quality)*, and removed by a **djoin**

5. CONCLUSIONS

Relations are perhaps the earliest form of persistent data type to be considered for embedding in a programming language [MERR77], [SCHM77]. This paper shows that, in conjunction with a suitably generalized relational algebra, they are widely useful, not only in data processing but in text and picture processing and logic programming. Developments of programming languages and systems on this basis is the objective of the Aldat project at McGill University. Our response to Michael Stonebraker's challenge, quoted at the beginning of this paper, is to say that Aldat provides the necessary extensions to the relational model for the next leap in database functionality.

188

6. ACKNOWLEDGEMENTS

I thank the Natural Science and Engineering Research Council of Canada and the Fonds formations de chercheurs et actions concertees du Quebec for support under grants NSERC A4365 and FCAC EQ2002 respectively.

REFERENCES

[ABCCM83] [AGOP85] [AITK85] [AM85] [BANC85] [BMS84] [BMW84] [BR84] [BUNE85] [CHIU82] [CM81] [CODD70] [CODD71a] [DATE83] [DHIL85] [DUEC83] [KNUT79] [MD85] [MERR77] [MERR84] [MERR85a] [REIT84] [SCHM77] [SS77] [VANR83] [ZW85]

Part III

Concurrency, Transactions and
Implementation of Persistence

Chapter 13

Linguistic Support for Atomic Data Types

William E. Weihl[1]
MIT Laboratory for Computer Science

1. INTRODUCTION

There are many applications in which the manipulation and preservation of long-lived, on-line data is of primary importance. Examples of such applications are banking systems, airline reservation systems, office automation systems, database systems, and various components of operating systems. A major issue in such systems is preserving the consistency of on-line data in the presence of concurrency and hardware failures. In this paper we consider how to define and implement data objects that help provide this consistency.

To support consistency it is helpful to make the activities that use and manipulate data *atomic*. Atomic activities are often referred to as *actions* or *transactions*; they were first identified in work on databases [DAVI73] [DAVI78] [EGLT76]. Atomic activities are characterized informally by two properties: serializability and recoverability. *Serializability* means that the concurrent execution of a group of activities is equivalent to some serial execution of the same activities. *Recoverability* means that each activity appears to be all-or-nothing: either it executes successfully to completion (in which case we say that it *commits*), or it has no effect on data shared with other activities (in which case we say that it *aborts*).

Nested transactions [DAVI73] [REED78] [MOSS81] [LYNC82] are useful for decomposing activities into smaller units. Nested transactions provide increased failure-tolerance: subtransactions of a transaction fail independently of each other and independently of the containing transaction. In addition, nested transactions can be used to run parts of the same activity concurrently, while ensuring that their concurrent execution is serializable. As discussed in [LISK82], nested transactions permit a simple implementation of a remote procedure call primitive with "at-most-once" semantics: a remote call is executed either zero or one times; partial and multiple executions cannot occur.

We have been exploring an approach in which atomicity is achieved through the shared data objects, which must be implemented in such a way that the activities using them appear to be atomic. Objects that provide appropriate synchronization and recovery are called *atomic objects* ; atomicity is guaranteed only when all objects shared by activities are atomic objects. By encapsulating the synchronization and recovery needed to support atomicity in the implementations of the shared objects, we can enhance modularity; in addition, by using information about the specifications of the shared objects, we can increase concurrency among activities.

[1] This work was supported in part by the Advanced Research Projects Agency of the Department of Defense, monitored by the Office of Naval Research under contract N00014-83-K-1025, and in part by the National Science Foundation under grant DCR-8203486.

Atomic objects are encapsulated within *atomic abstract data types*. An abstract data type consists of a set of objects and a set of primitive operations; the primitive operations are the only means of accessing and manipulating the objects [LZ74]. In addition, the operations of an atomic type ensure serializability and recoverability of activities using the type.

In our work on Argus [LISK83] [LS83] [WL85] we developed techniques for implementing atomic types, and designed linguistic constructs to support those techniques. In this paper we analyze the limitations of the techniques supported by Argus, and propose an alternative approach that has several advantages. We begin in Section 2 by discussing how to specify an atomic type, and how the specification of a type constrains the concurrency that may be permitted among activities using objects of the type. Then, in Section 3, we discuss the issues involved in implementing an atomic type. Next, in Section 4, we summarize and analyze the implementation techniques supported by Argus. The techniques supported by Argus are limited in several ways; in Section 5 we describe an alternative approach that avoids these limitations. Finally, in Section 6, we summarize our results and discuss further work.

2. SPECIFICATIONS AND ATOMICITY

A type's specification serves as a kind of contract between the implementor of the type and its users: the implementor guarantees that the type's behavior will obey the specification, and the users rely on this guarantee. Most importantly, however, the only assumptions that the users of a type can make about the type's behavior are those given in the type's specification. Thus, the implementor of a type has complete freedom in making implementation decisions, as long as the implementation satisfies the specification.

In writing specifications for atomic types, we have found it helpful to pin down the behavior of the operations, initially assuming no concurrency and no failures, and to deal with concurrency and failures later. In other words, we imagine that the objects exist in an environment in which all activities are executed sequentially, and in which activities never abort. We call this specification of a type's operations the *serial specification* of the type. The serial specifications of atomic types are particularly useful in reasoning about an activity that uses atomic objects. The atomicity of activities means that they are "interference-free," so we can reason about the partial correctness of an individual activity without considering the other activities that might be sharing objects with it [BR81]. This reasoning process is essentially the same as for sequential programs; the only information required about objects is how they behave in a sequential environment. The serial specification of an atomic type describes the assumptions that a single activity can make about the behavior of the type's objects, and serves to define the correct serial executions involving objects of the type.

Many different protocols can be used to ensure atomicity. However, if different types use different protocols, atomicity can be violated. To be able to ensure that the protocols used by different types are compatible, the specification of an atomic type must include some information about how the type manages concurrency and failures. We use the *behavioral specification* of a type to describe the concurrent executions permitted by the type. In [WEIH83] [WEIH84], we have explored constraints on the behavioral specifications of the types in a system that ensure that the types cooperate to guarantee atomicity. In the examples in this paper we will use one of those constraints, *dynamic atomicity*. A precise definition of dynamic atomicity can be found in [WEIH84]; in this paper we will use an informal definition, and focus on the issues involved in implementing atomic types.

Dynamic atomicity characterizes the behavior of objects implemented with protocols that determine the serialization order of activities dynamically based on the order in which the activities access the objects; for example, two-phase locking protocols [EGLT76] [KORT81] [BGL81] are dynamic atomic. Dynamic atomicity defines limits on the concurrency that can be permitted by an atomic type. These limits can be stated informally as follows: if the sequence of operations executed by two concurrent activities conflict, then some operation executed by one of the activities must be delayed until the other has completed (i.e., committed or aborted). Two activities are said to conflict if one has observed the effects of the other, or if one has invalidated the results of the other's operations.

As mentioned earlier, we permit activities to be nested. This means that an action can have subactivities, which may be concurrent. Regardless of its internal structure, however, an activity must appear atomic to activities outside it. If A is a subactivity of B, then the effects of A should not be visible outside B until both A and B commit. Thus, we have the following definition: we say that an activity A is *visible* to an activity B if A has committed up to the least common ancestor of A and B. In other words, A and each of its ancestors up to but not including the least common ancestor of A and B must have committed for A to be visible to B. If B invokes an operation that conflicts with one already executed by A, dynamic atomicity requires B's operation to be delayed until A is visible to B.

```
data type semiqueue[item: type] is create, enq, deq

% A semiqueue is like a bag (or multiset) of items.

create = proc () returns (semiqueue)
% Returns a new, empty semiqueue.

enq = proc (q: semiqueue, i: item)
% Adds i to q.

deq = proc (q: semiqueue) returns (item)
% If q is non-empty, chooses some element of q, removes it from q, and
returns it.
```

Fig. 2-1: Informal specification of the data type semiqueue.

For example, an (informal) serial specification of a data type *semiqueue* appears in Figure 2-1. A semiqueue is similar to a queue, except that enqueued items are not necessarily dequeued in first-in-first-out order. Instead, the *deq* operation makes a nondeterministic choice of an item to remove and return. If the semiqueue is empty, *deq* does nothing - in an actual concurrent execution, it waits for an item to be enqueued by some other activity. In the remainder of this paper we will use implementations of the semiqueue type to illustrate the use of different linguistic mechanisms.

Semiqueues are interesting because they place very few constraints on concurrency. For example, two *enq* operations do not conflict with each other, nor do an *enq* and a *deq* operation or two *deq* operations as long as they involve different elements. Thus, many different activities can *enq* concurrently, or *deq* concurrently. For example, if activity A has enqueued item X, B can dequeue X if A is visible to B. Once B dequeues X, however, a third action C cannot dequeue X unless B aborts. C could dequeue some other item Y if the activity that enqueued Y was visible to C.

3. IMPLEMENTATION ISSUES

Like an implementation of a data type in a sequential language, an implementation of an atomic type must define a representation for objects of the type, and must provide implementations for each operation of the type in terms of that representation. However, the implementation of an atomic type must also ensure appropriate synchronization and recovery for activities using objects of the type. As discussed above, the necessary synchronization and recovery are defined by the type's specification.

To provide synchronization and recovery for activities using objects of an atomic type, it is necessary to update the representation of objects as activities commit and abort. In Argus, the programmer relies on the system to update the representation. We call this an *implicit* approach. An alternative is for the programmer to supply code that is run when activities complete to update the representations of objects. We call this an *explicit* approach. In Section 4 we will focus on the implicit approach as supported by Argus. We will discuss an explicit approach and compare the two alternatives in Section 5.

In addition to providing appropriate synchronization and recovery for activities using objects of the type, an implementation of an atomic type must cope with internal concurrency and failures of the individual operations on the type. An operation invoked by an activity is not executed instantaneously: it may fail after completing only some of the steps described by its implementation. Operations invoked by concurrent activities may also run concurrently. Steps must be taken by the implementation of the type to manage concurrency and failures of operations.

4. ARGUS

In this section we discuss how atomic types can be implemented in Argus. The section is divided into three parts. In the first, we discuss the linguistic support in Argus for implementing atomic types. In the second, we present an example implementation. Finally, in the third, we discuss the strengths and weaknesses of the approach taken in Argus.

4.1 Linguistic Support

The mechanisms in Argus have two important characteristics. First, the names of activities are not accessible to user code. Second, no user code runs when activities complete. Instead, the programmer must rely on the system to update the representations of objects when activities complete. The linguistic support in Argus consists of several built-in atomic types, statements that use those types, and a mutual exclusion primitive.

The only processing done by the Argus system when an activity completes is to update locks and versions in the representation of each object of a built-in atomic type. Because of this, the programmer has to include some lower-level atomic objects (and ultimately at the lowest level some built-in atomic objects) in the representation of a user-defined atomic type. However, to implement types that permit highly concurrent use, the programmer must include non-atomic objects in addition to atomic objects. Some kind of synchronization and recovery is needed for these objects to cope with internal concurrency and failures; this is the purpose of the mutual exclusion primitive in Argus. The details of the Argus mechanisms are described below; a more complete description can be found in [LISK83], and motivation and explanation of their use can be found in [WL85].

4.1.1 The Type Generator Atomic_variant

Argus provides several built-in atomic types and type generators. Of particular interest to us in our examples is the built-in type generator **atomic_variant**. The serial specification of atomic variants is essentially that of variants in CLU [LISK81]: a variant type specification consists of a list of tags and associated types. The state of a variant object consists of a tag and a value; if the current tag of a variant is t, then the type of the current value of the variant is the type associated with t. For each tag t in the type specification, there are four operations: *make_t, change_t, is_t,* and *valuet*. *Make_t* takes one argument of the type associated with the tag t, and returns a new variant object whose tag is t and whose value is the argument of the operation. *Change_t* takes two arguments, a variant and an object of the type associated with t, and changes the state of the variant so that its tag is t and its value is the second argument of the operation. *Is_t* takes one variant argument, and returns **true** if and only if the tag of the variant is t. Finally, *value_t* takes one variant argument; if the tag of the variant is t then it returns the current value, and otherwise it signals *wrong_tag*. When only these operations are used, atomic variants are dynamic atomic.

Atomic variants are used in two ways: in conjunction with other atomic objects, to make activities atomic; and in conjunction with non-atomic objects, as part of the representation of a user-defined atomic object. In the latter case, it may be possible for an activity to gain access to an atomic variant created by another activity that has aborted or is still active. For example, one activity might create an atomic variant and insert it in a (non-atomic) array; a concurrent activity with access to the array could then access the newly created atomic variant. Thus, the *make_t* operations, which create new atomic variant objects, require special consideration: we must define what happens to an atomic variant when its creator aborts, and whether concurrent activities can use an atomic variant before the object's creator has completed.

Argus defines the *make_t* operation to create a new atomic variant object whose tag is t and whose value is the argument to the operation; this state is the object's "base" state, and the object will continue to exist in this state even if the creating activity aborts. (An alternative is to have the object "disappear" when its creator aborts; our experience indicates that this leads to awkward and complex programs.)

Concurrent use of a newly created atomic variant is limited as follows: The operations on atomic variants are classified as readers and writers. Synchronization of activities using an atomic variant object is done with read and write locks. The usual locking rules apply: Any number of activities can hold read locks simultaneously, but if one activity holds a write lock then no other activity can hold a read lock or a write lock. *Make_t, is_t,* and *value_t* are all readers, and *change_t* is a writer; readers acquire read locks when executed, and writers acquire write locks. By having the activity that creates an atomic variant retain a read lock, we ensure that the activity will not observe concurrent use of the object by other activities.

4.1.2 The Tagtest Statement

In an implementation of a user-defined atomic type, it is convenient for an activity to be able to test whether it would have to wait if it were to invoke a particular operation on an atomic variant. Argus provides the **tagtest** statement as structured support for testing and setting locks on atomic variants. The use of the **tagtest** statement can violate atomicity, since it permits an activity to observe the presence of concurrent activities. However, it appears to be necessary for implementing user-defined atomic types using an implicit program structure.

A **tagtest** statement has the following form:[2]

```
tagtest expression
    atag_arm { atag_arm }
    [ others : body ]
    end
where
    atag_arm ::= tag_kind name, ... [ (idn:type_spec) ] : body
    tag_kind ::= tag
               | wtag
```

The expression must evaluate to an atomic variant object. If a read lock could be obtained on the atomic variant object by the activity executing the statement, then the tag of the object is matched against the names on the atag_arms; if a matching name is found, then the tag_kind on the arm is considered.

If the tag_kind is **tag**, a read lock is obtained on the object and the match is complete. If the tag_kind is **wtag** and the activity can obtain a write lock on the object, then a write lock is obtained and the match is complete. In all other cases the match is incomplete.

If a complete match is not found, or the activity could not obtain a read lock, then the body in the **others** arm, if present, is executed; if there is no **others** arm, the **tagtest** statement terminates.

When a complete match is found, if a declaration (*idn* : *type_spec*) appears on the matching arm, the value component of the object is assigned to the local variable *idn* . The body on the matching arm is then executed; *idn* , if declared, is defined only in that body. The entire matching process, including testing and acquisition of locks, is indivisible.

4.1.3..Mutual Exclusion

Argus provides the built-in type generator **mutex** and the **seize** statement to enable implementations to cope with concurrency among executions of operations. **Mutex** and **seize** can be used to ensure mutual exclusion among regions of code executed by concurrent operations; thus, for example, implementations can prevent interference among concurrently executing operations by forcing them to run serially.

Mutex objects are mutable containers for information. The type generator **mutex** has a single parameter, which is the type of the contained object. Mutex types provide operations to create and decompose mutex objects. The *create* operation takes a single argument of the parameter type and creates a new mutex object containing the argument object. The *get_value* operation extracts and returns the contained object from its mutex argument; following the conventions of Argus, the expression "**mutex**[t]$get_value(m)" is usually written "m.value".

---

2 We use an extended BNF for syntactic descriptions, with the following conventions: | is used to separate alternatives; "[*a*]" denotes an optional *a*; "{*a*}" denotes a sequence of zero or more *a*'s; and "*a*, ..." denotes a list of one or more *a*'s separated by commas.

Mutexes are used primarily to provide mutual exclusion on non-atomic shared data. Argus provides the **seize** statement, which allows a sequence of statements to be executed by an activity while the activity is in exclusive possession of a mutex object. The **seize** statement has the following form:

$$\text{seize expression } \textbf{do} \text{ body } \textbf{end}$$

An implementation of an operation often has a precondition that must be true before the operation can be executed. Sometimes it is necessary to gain possession of mutex objects simply to test the precondition; if the precondition is false, the operation must wait. It is important to release possession of mutex objects while waiting, particularly if some other operation that requires possession of the mutex objects must be executed for the precondition to become true. Argus provides the **pause** statement for this purpose. It may be executed only inside the body of a **seize** statement. When a process executes **pause**, the mutex object seized by the closest enclosing **seize** statement is released, and the process is blocked for a system-determined time. When the process is unblocked, it regains possession of the mutex object released by the **pause** statement, waiting if necessary, and then continues execution with the statement following the **pause**.

The expression must evaluate to a mutex object. After evaluating the expression, the executing activity attempts to gain *possession* of the resulting mutex object. Only one activity may have possession of a mutex object at one time; thus, an activity may be forced to wait when it attempts to gain possession. Once the activity gains possession, the body of the **seize** statement is executed. Termination of the body causes possession of the mutex object to be released. If several processes are waiting for possession of the same mutex object, possession will be awarded *fairly* , in the sense that as long as no process retains possession forever, every waiting process will eventually gain possession.

Mutex objects are also used in Argus to protect against the effects of failures while an activity is executing an operation on an atomic object. An activity in Argus never fails while in possession of a mutex object unless the containing guardian crashes, causing the activity to be aborted. When such a crash occurs, the states of the objects in the guardian are restored from stable storage. Since a mutex object is not copied to stable storage while it is possessed by an activity, the state of a mutex object recovered from stable storage will not show the effects of the activity aborted by the crash. A more complete discussion of the interactions between implementations of atomic types and stable storage in Argus can be found in [WL85]. In this section and the next we will assume that an activity in possession of a mutex object executes the body of the **seize** statement correctly, and does not abort until after it has released possession of the mutex.

4.2 Implementation of the Semiqueue Type

Our first example is an implementation of the *semiqueue* type. An informal specification of the semiqueue type was presented in Figure 2-1. As discussed in Section 2, semiqueues place few constraints on concurrency. Many different activities can *enq* concurrently, or *deq* concurrently. Furthermore, one activity can *enq* while another *deq*'s, provided only that the *deq* does not return the newly *enq*'d item. Similarly, activities can *deq* concurrently as long as the *deq* operations return different items.

```
semiqueue = cluster[item: type] is create, enq, deq

qitem = atomic_variant[enqueued: item,
                       dequeued: null]
buffer = array[qitem]
rep = mutex[buffer]

create = proc () returns (cvt)
  return(rep$create(buffer$new()))
  end create

enq = proc (q: cvt, i: item)
  qi: qitem := qitem$make_dequeued(nil) % dequeued if activity aborts
  qitem$change_enqueued(qi, i)          % enqueued if activity commits
  seize q do
    b: buffer := q.value
    buffer$addh(b, qi)                         % add new qitem to buffer
    end
  end enq

deq = proc (q: cvt) returns (item)
  cleanup(q)       % cleanup should be called less frequently
  seize q do
    b: buffer := q.value
    while true do
      for qi: qitem in buffer$elements(b) do
        tagtest qi % see if item can be dequeued by this activity
          wtag enqueued (i: item): qitem$change_dequeued(qi,nil)
                                   return(i)
          end
        end
      pause
      end
    end
  end deq

cleanup = proc (q: rep)
  enter topaction        % start an independent activity
    seize q do
      b: buffer := q.value
      for qi: qitem in buffer$elements(b) do
        % remove only qitems in the dequeued state
        tagtest qi
          tag dequeued:  buffer$reml(b)
          others: return
          end
        end
      end
    end
  end cleanup

end semiqueue
```

Fig. 4-1: Implicit implementation of the data type semiqueue.

An implementation of the semiqueue data type appears in Figure 4-1. The implementation is written as a *cluster* , the module provided by Argus and CLU [LISK81] for implementing an abstract data type. A cluster consists of a *header* listing the names of the operations provided by the type; some *equates*, defining abbreviations for types, in particular the abbreviation **rep**, which defines the representation type for the implementation; and *operations* , which are simply procedures implementing each of the operations provided by the type. The keyword **cvt** used as the type of an argument or a result of an operation indicates that the indicated argument or result is viewed as an object of the representation type inside the cluster, and as an object of the type defined by the cluster outside the cluster.

The plan of this implementation is to keep the enqueued items in a regular (non-atomic) array. This array can be used by concurrent activities, but it is enclosed in a mutex object to control internal concurrency. All modification and reading of the array occurs inside a **seize** statement on this containing mutex object. Arrays in Argus are like arrays in most languages, except that they are extensible: they can grow and shrink dynamically at both ends. The semiqueue implementation uses four operations on arrays: *new* , which creates and returns a new empty array (one with no elements); *addh* , which appends a new element on the high end of its array argument; *reml* , which removes and returns an element from the low end of its array argument, signalling *bounds* if the array is empty; and *elements* , which is an iterator that yields the elements of its array argument in order for the lowest index to the highest.

To determine the status of each item in the array, we associate with each item an atomic object that tells the status of activities that inserted or deleted that item. For this purpose we use the built-in atomic type **atomic_variant** (described in the previous section).

The semiqueue operations are implemented as follows: The *create* operation simply creates a new empty array and places it inside a new mutex object. The *enq* operation associates a new atomic variant object with the argument item; this atomic variant will have tag "enqueued" if the calling activity commits later, and tag "dequeued" if it aborts. Then *enq* seizes the mutex and adds the new atomic variant to the contained array.

The *deq* operation seizes the mutex and then searches the array for an item it can dequeue. Such an item must have been enqueued by an activity that is visible to the activity executing *deq*, and must not have been dequeued by an active or committed activity. Thus, the *deq* operation searches for an atomic variant that has tag "enqueued" and on which the activity that called *deq* can obtain a write lock. Upon finding such an atomic variant, the contained item is selected and returned after changing the variant's tag to "dequeued". If no suitable atomic variant is found, **pause** is executed (releasing the mutex) and later the search is retried.

Proper synchronization of activities using a semiqueue is achieved by using the *qitems* in the buffer. An *enq* operation need not wait for any other activity to complete: it simply creates a new *qitem* and adds it to the array. Of course, it may have to wait for another operation to release the mutex object before adding the *qitem* to the array, but this delay should be relatively short. A *deq* must wait until some activity that executed an *enq* operation commits relative to the activity that invoked *deq*; thus it searches for a *qitem* with tag "enqueued" that it can write.

The *qitems* are also used to achieve proper recovery for activities using a semiqueue. Since the array in the mutex is not atomic, changes to the array made by activities that abort later are not undone. This means that a *deq* operation cannot simply remove a *qitem* from the array, since this change could not be undone if the calling activity later aborted. Instead, a *deq* operation changes the state of a *qitem*; the atomicity of *qitems* ensures proper recovery for this modification. If the calling activity later commits to the top level, the *qitem* will have tag "dequeued" permanently. Such *qitems*, which are also generated by *enq* operations called by activities that later abort, have no effect on later operations. Leaving them in the array wastes storage, so the internal procedure *cleanup*, called by *deq*, removes them from the low end of the array. (Of course, a more realistic implementation would call *cleanup* only occasionally.)

Note that *cleanup* cannot run in the calling activity: if the calling activity had previously executed a *deq* operation, that *deq* would be visible to a later operation executed by the same activity. Instead, *cleanup* runs as an independent activity. This activity will only be able to lock *qitems* that are not being used by any active activities; thus it will not remove any *qitems* that could affect later operations.

4.3 Remarks

The implementation of the semiqueue type in the previous section illustrates the general strategy used to implement a user-defined atomic type in Argus. The representation of a user-defined atomic type typically consists of a mutex object containing a non-atomic collection (e.g., an array) of atomic objects (typically atomic variants). Greater concurrency among activities using the type is achieved by introducing atomic objects only at the lowest level of the representation. The internal *cleanup* routine, used to discard atomic objects that are no longer needed from the representation, is also typical of implementations of user-defined atomic types in Argus.

The implementation in the previous section also illustrates a number of limitations of the expressive power of the implicit approach supported by Argus.

First, the implementation of *deq* is relatively inefficient, since in the worst case it takes time proportional to the size of the representation of the semiqueue. There is no obvious way to improve the efficiency of this implementation: the activity that executes an operation is implicit, so there is no way to structure the representation of an object based on the activities that enqueued or dequeued an item.

Second, scheduling of *deq* operations is accomplished using busy-waiting. The system has very little information on which to base scheduling decisions, implying that an activity with a pending operation is likely to be awakened when it is unable to complete the operation, and also that an activity may be unlikely to be awakened very soon after the precondition for the operation becomes true.

Finally, the programmer has no control over when the representation of an object gets updated by the system as activities commit and abort. In Argus, the system updates built-in atomic objects automatically when activities complete, and does so at arbitrary times. The following example illustrates the problems that can result. Suppose that we want to implement the semiqueue type with the following additional constraint: If there is only one dequeuing activity at a time, and dequeuing activities do not abort, then items enqueued by a single activity should be dequeued in the order in which they were enqueued. This constraint is not satisfied by the implementation presented above. Suppose that activity A has enqueued two items, X and Y, in that order, and that activity B starts to execute a *deq* operation. If A commits after B has examined the first *qitem* (containing X) in the representation of the semiqueue and before B has examined the second, the *deq* operation will return Y.

It seems impossible to modify the implementation of semiqueue presented above to satisfy this additional constraint, given the semantics of Argus: if *deq* has found an item that can be dequeued, there is no way to tell whether some other item was enqueued by the same activity. If we impose the additional restriction on the system that commits and aborts appear to be instantaneous (so if an activity has committed at one atomic variant then it has committed at any others that it touched), then we can modify *deq* to search backwards through the representation and to return the last available item that it finds. (Or to search forwards until it finds one available item, and then to search backwards from there.) The resulting implementation satisfies the additional constraint on semiqueues. However, it could be expensive to make commits and aborts appear instantaneous in a distributed system.

This example illustrates that the programmer does not have complete control over all events that affect the representation of an object. Commit and abort events involving lower-level objects are controlled by the system, and can occur asynchronously. This asynchrony is visible to the programmer, and can affect the correctness of an implementation.

In the next section we will present an alternative approach that avoids the problems discussed above.

5. SUPPORT FOR AN EXPLICIT APPROACH

In this section we discuss an explicit approach for implementing atomic types, in which the programmer supplies code that is run when activities complete to update the representations of objects. We begin in Section 5.1 by describing extensions to Argus to support the explicit approach. Then, in Section 5.2, we present an example implementation to illustrate the approach. Finally, in Section 5.3, we compare the implicit and explicit approaches.

5.1..Linguistic Support

We present the language constructs as additions to Argus. We do not intend this to be a complete language proposal; rather, it is a vehicle for presenting examples using an explicit approach. The examples will serve both to illustrate how implementations of atomic types can be constructed using an explicit approach, and as a basis for comparing the explicit and implicit approaches.

We extend Argus in three ways. First, we add a new built-in data type, **aid**, to represent names of activities. Second, we extend the existing module for implementing abstract data types (the **cluster**) to provide the implementation of an operation with access to the name of the activity that invoked it, and to permit easy identification to the system of the code to be run when activities commit and abort. Third, we add a queuing and signalling mechanism designed to support efficient synchronization of nested activities.

```
data type aid is parent, ancestors, proper_ancestors, top, equal

parent = proc (a: aid) returns (aid) signals (top)
% Returns a's parent; signals top if a has no parent.

ancestors = iter (a: aid) yields (aid)
% Yields the ancestors of a, including a itself, in root-to-leaf order.

proper_ancestors = iter (a: aid) yields (aid)
% Yields the proper ancestors of a (i.e., not including a itself) in root-to-leaf order.
```

```
top = proc (a: aid) returns (bool)
```
% Returns **true** if *a* is a top-level activity; otherwise returns false.

```
equal = proc (a1, a2: aid) returns (true)
```
% Returns **true** if *a1* and *a2* name the same activity; otherwise returns false.

Fig. 5-1: Informal specification of the data type **aid**

An informal specification of the type **aid** appears in Figure 5-1. Note that no operations are provided to create new aids. While we will allow an implementation of an atomic type explicit access to the aid of an activity that invokes one of the type's operations, we follow Argus in implicitly associating aids with processes. Thus, the system creates new aids automatically whenever an existing activity executes the **enter** statement to create subactivities or nested top-level activities, and associates the new aids with the corresponding processes. Also note that the set of operations provided by the type **aid** is not complete; we have included only those that we need for our examples, and expect that others would be needed for general use.

We extend clusters in two ways. First, a routine in a cluster that implements an operation of the defined type can have two interface specifications. The *external* specification corresponds to the interface specification of the operation in the type's specification. The *internal* specification differs from the external specification in that it has an additional implicit argument. This implicit argument must appear as the first argument in the argument list of the routine, and has type **aid**. Thus, for example, an operation with external specification

```
op = proc (x1: t1, ...) ...
```

might have internal specification

```
op = proc (a: aid, x1: t1, ...) ...
```

The identifier used to declare the implicit argument may be chosen at the convenience of the programmer. When a routine with distinct internal and external specifications is invoked, the implicit argument is assigned the value of the **aid** of the invoking activity, and the other arguments are assigned the values of the corresponding actuals.

Second, a cluster may supply two additional operations, called **commit** and **abort**. We call these special operations *completion operations*. Their interfaces are as follows:

```
commit = proc (a: aid, x: rep)
```

```
abort = proc (a: aid, x: rep)
```

These operations are intended to be called by the system (say, with arguments *a* and *x*) when an activity *a* that used the object represented by *x* commits or aborts. To let the system know that an activity has used an object, the routines inside a cluster may call the special procedure **register**, which has the following interface:

```
register = proc (a: aid, x: rep) signals (completed)
```

An invocation of **register** will signal *completed* if the activity named by the first argument has already committed or aborted. Otherwise the invocation will return. (This is not an expected occurrence, and probably represents a programming error.) Sometime after the activity *a* completes, the system will invoke the appropriate completion operation defined in the cluster (**commit** if the activity commits, **abort** if the activity aborts) with arguments *a* and *x*. If the completion operation signals *failure*, the system will try again at some future time. If a committing activity is a subactivity of another activity, and the invocation of the **commit** operation terminates normally, the system will also register the activity's parent on the same object; thus, when the parent completes, the appropriate completion operation will again be invoked. (Similarly, if the parent commits and its **commit** operation terminates normally, the system will register its parent on the same object, and so on until a top-level activity is reached.)

We will use the **mutex** type and the **seize** statement in Argus to cope with internal concurrency. As in the previous section, we assume that activities do not fail when in possession of a mutex object.

```
data type action_queue is create, notify, wake, empty
```

% A process can add itself to an **action_queue** by executing the **block** statement,
% specifying an **aid** on whose behalf it wishes to wait. A process on an **action_queue** is
% in one of two states: *asleep* or *waiting*. A *waiting* process will be unblocked as soon
% as it can regain possession of the mutex object released when it blocked. An
% **action_queue** is *empty* if and only if no processes, asleep or waiting, are blocked on it.

```
create = proc () returns (action_queue)
```
% Returns a new, empty action queue.

```
notify = proc (q: action_queue, a: aid)
```
% Changes all *asleep* processes on *q* waiting on behalf of siblings of *a* or their
% descendants to *waiting*; all top-level activities are considered to be siblings of a
% top-level activity.

```
wake = proc (q: action_queue)
```
% Changes all *asleep* processes on *q* to *waiting*.

```
empty = proc (q: action_queue) returns (bool)
```
% Returns **false** if any process, *asleep* or *waiting*, is blocked on *q*; otherwise returns **true**.

Fig. 5-2: Informal specification of the data type **action_queue**

Finally, we add the built-in data type **action_queue** to allow operations to wait for necessary preconditions. An informal specification of the operations provided by **action_queue** appears in Figure 5-2. (The *empty* operation on action queues, used to test whether any activities are blocked on a queue, is not used in the example in this paper. We show it for completeness because we have found it useful in other examples.) We also provide the **block** statement to allow an activity to wait on an **action_queue**. The **block** statement has the following form:

```
block expr1 on expr2
```

The first expression must evaluate to an **aid**, and the second to an **action_queue**. The **block** statement can appear only within a **seize** statement. When executed, it blocks the executing process on the specified **action_queue** on behalf of the specified **aid**, and releases the mutex object seized by the closest enclosing **seize** statement. The **block** statement is a statement, rather than an operation on action queues, so that the programmer need not be concerned with releasing and reacquiring possession of the mutex object seized by the enclosing **seize** statement. The types **aid** and **action_queue** are built-in because they are used by the **block** statement.

A process blocked on an **action_queue** is in one of two states: *asleep* or *waiting*. When a process executes a **block** statement, it is initially *asleep*. An *asleep* process on an **action_queue** may change to the *waiting* state when some other process executes the *notify* or *wake* operation on the **action_queue**. A process in the *waiting* state attempts to regain possession of the mutex object that was released when the process blocked, and is unblocked as soon as it regains possession.

5.2 Implementation of the Semiqueue Type

data type log[t: **type**] **is** create, fetch, store, delete, ancestors

% A *log[t]* object maps **aids** to *t* objects.

create = **proc** () **returns** (log[t])
% Returns a new, empty log.

fetch = **proc** (l: log[t], a: **aid**) **returns** (t) **signals** (not_found)
% Returns the *t* object associated with *a* in *l*, signalling *not_found* if *a* is not bound in *l*.

store = **proc** (l: log[t], a: **aid**, x: t)
% Binds *a* to *x* in *l*.

delete = **proc** (l: log[t], a: **aid**)
% Unbinds *a* in *l*.

root2leaf = **iter** (l: log[t], a: **aid**) **yields** (aid, t)
% Yields each ancestor of *a* (including *a* itself) with its associated binding, if it is bound
% in *l*; items are yielded in root-to-leaf order.

leaf2root = **iter** (l: log[t], a: **aid**) **yields** (aid, t)
% Yields each ancestor of *a* (including *a* itself) with its associated binding, if it is bound
% in *l*; items are yielded in leaf-to-root order.

Fig. 5-3: Informal specification of the data type log

In this section we present an implementation of the semiqueue type using the linguistic constructs described in the previous section. An informal specification of semiqueues appeared in Figure 2-1. The implementation appears in Figure 5-4. It uses the type generator *log*; an informal specification of logs appears in Figure 5-3. A log can be used to store and retrieve information based on an **aid**, making it easy to keep track of the history of each activity.

The representation of a semiqueue consists of three components enclosed in a mutex object. The components are: *committed*, which represents the items known to be in the semiqueue (they have been enqueued by activities that have committed to the top-level, and they have not been dequeued); *logs*, which is a collection of summary information about the operations executed by active activities; and *pending*, which is an activity queue used for blocking *deq* operations that cannot find an item to dequeue. The mutex object is used to prevent interference among concurrently executing operations on the same semiqueue by forcing them to run serially.

```
semiqueue = cluster [item: type] is create, enq, deq

undo = record[i: item,                    % Item returned by deq op.
              deleted_from: elist]        % elist it was removed from.

elist = array[item]                       % Intentions list for enq's, and list
                                          % of fully committed enq'd items.

dlist = array[undo]                       % Undo log for deq's

summary = record[enq: elist,              % summary of operations executed

......          deq: dlist]                % by a single activity

aq = action_queue

components = record[committed: elist,     % committed items in semiqueue
                    logs: log[summary],   % summaries for all activities
                    pending: aq]          % pending deq ops.

rep = mutex[components]

% rep invariant:
%    for each activity a, and for each undo record u in rep.value.logs[a].deq, either
%    u.deleted_from = rep.value.commited, or there exists a proper ancestor a' of a such
%    that u.deleted_from = rep.value[a'].enq

create = proc () returns (cvt)
  return(rep$create(components${committed: elist$new(),
                               logs: log[summary]$create(),
                               pending: aq$create()}))
  end create

% external spec: enq = proc (q:cvt, i: item)
enq = proc(a: aid, q: cvt, i: item)
  seize q do
    s: summary := find_log(q.value.logs, a)
    elist$addh(s.enq, i)
    register(a, q)
    end
  end enq

% external spec: deq = proc (q:cvt) returns (item)
deq = proc(a: aid, q: cvt) returns (item)
  seize q do
    while true do
      visible: elist := find_elist(q.value,a)
        except when none: block a on q.value.pending % Pause
                          continue    % and retry at the start of the loop
          end
      u: undo := undo${i: elist$reml(visible), deleted_from: visible}
      s: summary := find_log(q.value.logs.a)
      if u.deleted_from ~= s.enq
        then dlist$addh(s.deq, u) end % else operations cancel each other
      register(a, q)
      return(u.i)
      end
    end
  end deq
```

```
% gets the summary for a from logs.
find_log = proc (logs: log[summary], a: aid) returns (summary)
  return(log[summary]$fetch(logs,a))
    except when not_found: s: summary := summary${enq: elist$new(),
                                                 deq: dlist$new()}
                      log[summary]$store(logs,a,s)
                          return(s)
        end
  end find_log
```

% finds a non-empty elist visible to a (i.e., either committed, or belonging to an ancestor
% of a). If q.committed is non-empty it is returned. Signals if no non-empty elist is
% found

```
find_elist = proc (c: components, a: aid) returns (elist)
                                          signals (none)
  if elist$size(c.committed) ~= 0 then return(c.committed) end
  for anc: aid, s: summary in log[summary]$root2leaf(c.logs, a) do
    if elist$size(s.enq) ~= 0 then return(s.enq) end
    end
  signal none
  end find_elist

commit = proc (a: aid, q: rep)
  seize q do
    qv: components := q.value
    l: log[summary] := qv.logs
    as: summary := log[summary]$fetch(l,a)
      except when not_found: return end
    log[summary]$delete(l, a)
    ps: summary := find_log(l, aid$parent(a))
      except when top: merge_enq(qv.committed, as.enq)
                       if elist$size(as.enq) ~= 0
                       then aq$wake(qv.pending)
                         end
                       return
        end
    merge_enq(ps.enq, as.enq)
    merge_deq(ps, as.deq)
    if elist$size(as.enq) ~= 0 then aq$notify(qv.pending, a) end
    end
  end commit

  abort = proc (a: aid, q: rep)
    seize q do
      as: summary := log[summary]$fetch(q.value.logs,a)
        except when not_found: return end
      for u: undo in dlist$elements(as.deq) do
        elist$addh(u.deleted_from, u.i)
        end
      if dlist$size(as.deq) ~= 0 then aq$wake(q.value.pending) end
      end
    end abort
```

% appends (in order) items in from onto to.
```
merge_enq = proc (to: elist, from: elist)
  for i: item in elist$elements(from) do
    elist$addh(to, i)
    end
  end merge_enq
```

```
% appends (in order) undos in from onto to.deq, ignoring those undos u
% with u.deleted_from = to.enq (such deq's have committed to the
% level of the corresponding enq, so both operations can be forgotten).
merge_deq = proc (to: summary, from: dlist)
    for u: undo in dlist$elements(from) do
      if u.deleted_from ~= to.enq
        then dlist$addh(to.enq, u)
        end
      end
    end merge_deq

end semiqueue
```

Fig. 5-4: Explicit implementation of the data type semiqueue.

The summary for an activity consists of two parts: *enq*, which represents the items enqueued by the activity (or its committed descendants) and not subsequently dequeued; and *deq*, which represents the items dequeued by the activity (or its committed descendants), and contains sufficient information to be able to "undo" the *deq* operations if the activity aborts.

The implementation of *enq* works as follows: it first finds the summary record for the invoking activity by calling the internal procedure *find_log*. It then adds the item to be enqueued to the list of items enqueued by the activity, registers the invoking activity and the semiqueue object (so the appropriate completion operation will be invoked by the system when the activity completes), and returns. The mechanism used for *enq* operations is like an intentions list: a record of the operation is kept, but the operation only becomes visible to the activity's siblings when the activity commits. If the activity aborts, the record of the operation is discarded.

The implementation of *deq* is more complex: it first looks for an item to dequeue by calling the internal procedure *find_elist*. *find_elist* searches the committed items and the intentions lists for the calling activity and its ancestors, looking for a non-empty list. The lists searched by *find_elist* contain the enqueued items that are visible to the calling activity and that have not yet been dequeued. If no list is found by *find_elist*, *deq* blocks on the activity queue in the representation of the semiqueue, and tries again when some activity that used the semiqueue becomes visible to the calling activity. If a non-empty list is found by *find_elist*, *deq* removes the first item from the list and creates an undo record containing that item and the list. Next, if the item dequeued was not enqueued by the invoking activity (or one of its committed descendants), the undo record is added to the summary information for the activity. (If the same activity enqueued and then dequeued an item, there is no need to remember either operation; the net effect on the semiqueue will be the same regardless of whether the activity commits or aborts.) Finally, the invoking activity and the semiqueue are registered, and the item to be dequeued is returned.

The mechanism used for *deq* operations is like an undo log: if the invoking activity aborts, the information in the undo records is used to put the item back in the list from which it was removed, effectively "undoing" the operation.

The **abort** routine is simple: it undoes the *deq* operations executed by the aborting activity and its committed descendants by putting the dequeued items back in the lists from which they were removed, and then discards the summary (including the intentions list of enqueued items) for the aborting activity. Finally, if any *deq* operations were undone, the **abort** routine unblocks any pending *deq* operations, since the items that were returned from the aborting activity's undo list might now be visible to the pending operations.

The **commit** routine merges the summary for the committing activity with that for its parent. The activity's list of enqueued items is simply appended to its parent's list. The undo information for *deq* operations is similarly appended, except that records for items enqueued by the parent are discarded from the undo log. (The operations effectively cancel each other in this case.) In the case of a committing top-level activity, the tentatively enqueued items are appended to the list of committed items, and the undo log is discarded. Finally, if any items enqueued by the committing activity were added to its parent's intentions list, the **commit** routine unblocks pending *deq* operations invoked by activities to which the committing activity is now visible.

When a blocked *deq* operation is unblocked, there is no guarantee that it will be able to execute. For example, there may be several pending *deq* operations unblocked by the same completing activity, but it is possible that only one will actually be able to dequeue an item. Thus, the *deq* operation loops after blocking, and may block again if it still finds no items available for it to dequeue.

5.2. Remarks

In this section we discuss the relative merits of the implicit and explicit approaches. We begin in Section 5.3.1 by comparing and evaluating the two approaches. Then, in Section 5.3.2, we discuss related work.

5.3.1. Comparison and Evaluation

The explicit implementation of the semiqueue type has a number of advantages over the implicit implementation. First, the implementation of the *deq* operation in the explicit implementation is more efficient than in the implicit implementation. In the implicit implementation the *deq* operation takes time proportional to the size of the representation of its semiqueue argument (in the worst case); in the explicit implementation the *deq* operation takes time proportional to the number of ancestors of the calling activity (again in the worst case). This difference arises because the explicit implementation has access to the names of invoking activities.

Second, the semiqueue type cannot be implemented in Argus to satisfy additional properties, like the restriction that items enqueued by a single activity be dequeued in the order in which they are enqueued; the explicit mechanism presented here does not suffer from this limitation. The problem with Argus is that the system does not update the states of all built-in atomic objects instantaneously, so it is possible for one activity to see that another activity has committed at one built-in object, and later to see that the activity still holds a lock on another built-in object. It is possible to change the semantics of Argus to avoid this problem. However, it might be expensive for the system to guarantee that commits and aborts appear instantaneous in a distributed system. In addition, even with this additional guarantee, it can be difficult to implement the "fifo"-like restriction on semiqueues. The reader can easily verify that the explicit implementation above satisfies this restriction.

Third, the implicit implementation of the semiqueue type uses busy-waiting to schedule *deq* operations, while the explicit implementation uses a signalling mechanism. The signalling mechanism can be significantly more efficient than busy-waiting: in the implicit implementation, the system has very little information on which to base scheduling decisions, and is quite likely to awaken a pending operation when the operation cannot proceed, and not to awaken a pending operation when in fact it can proceed. In the explicit implementation, a pending *deq* operation is awakened only if an activity that enqueued some items becomes visible to the activity waiting to *deq*, or if an activity that dequeued some items aborts, making those items available for other activities to dequeue.

Finally, the implicit implementation requires a "cleanup" routine to keep the size of the representation of a semiqueue from increasing forever; the explicit implementation accomplishes the same effect with the user-supplied completion operations, and does it more efficiently since it is possible to tell from the arguments to the completion operation exactly what information needs to be deleted from the representation and when it needs to be deleted.

Achieving the flexibility in structuring representations provided by the explicit approach described above appears to require that operations have access to the names of invoking activities, but does not require the use of user-supplied completion operations. Instead, we could provide a "query" operation on **aid** objects, permitting an implementation to find out from the system whether a given activity is still active, and if not, whether it has committed or aborted. An implementation could periodically check the status of active activities, and update the representation of an object appropriately when told that an activity has completed. In this way we could achieve much of the effect of user-supplied completion operations, without having them executed automatically by the system.

While user-supplied completion operations are not needed for flexibility in structuring representations, they may be necessary for achieving efficient scheduling of operations. The implicit approach in Argus uses busy-waiting to schedule operations; as illustrated by the implementations of the semiqueue type, the queuing mechanism in the explicit approach presented here permits much more control over scheduling of operations. The mechanism relies on user-supplied completion operations to awaken blocked operations explicitly. It is not clear whether comparable control over scheduling can be achieved without user-supplied completion operations.

The explicit implementations that we have studied are typically more complex than implicit implementations of the same data types. Some of this complexity arises from the need to represent explicitly certain kinds of information that are handled automatically by the system in the implicit approach. For example, in Argus the system keeps track of the built-in atomic objects used by activities. In an explicit approach, the programmer must write code to keep track of the histories of activities. In the examples we have studied for which the mechanisms provided by Argus are particularly well-suited, the extra code needed in an explicit implementation can be substantial.

Some of the complexity, however, arises out of the desire for more efficient and more concurrent implementations. Some of the information that is handled automatically by the system in the implicit approach may be more efficiently used when structured differently (e.g., compare the two semiqueue implementations). As discussed above, the explicit approach, by permitting implementations to access the names of invoking activities, provides control over this structure, while the implicit approach does not. In addition, the management of action queues introduces complexity, but also permits much more efficient scheduling.

More work may be required of the programmer in the explicit approach, since it may be necessary to supply completion operations. However, in the implicit approach, it is frequently necessary to provide an internal operation that compacts the representations of objects (e.g., the *cleanup* routine in the implicit implementation of the semiqueue type); this kind of internal garbage collection is not needed in explicit implementations.

The use of mutual exclusion to control internal concurrency is just a simple way of making the implementations of operations "atomic," by forcing them to run serially. A more general approach is to use atomic actions, making each execution of an operation a top-level activity. This approach appears viable, however, only if implementations have access to the names of invoking activities. In an implicit structure, certain steps in the implementation of an operation (e.g., operations on lower-level atomic objects) must be executed on behalf of the invoking activity, yet there is no way for a process to act on behalf of two activities at the same time. Even in an explicit structure, this approach may have difficulties. It is not clear whether a queuing mechanism comparable to our **action_queue**'s can be designed to work with top-level activities instead of with **seize** statements.

In summary, the expressive power of the implicit approach is limited in several ways. However, implementations using an explicit approach appear more complex. It is not clear whether an intermediate approach can be found that provides the efficiency of the explicit approach but avoids some of the complexity.

5.3.2. Related Work

Schwarz and Spector [SS84] [SCHW84] at CMU have been studying how to build "atomic objects." They focus only on locking implementations, however, and appear to suggest that the system should manage "lock tables" automatically. They do not describe in detail how the programmer can describe the set of lock modes and their conflict relationships to the system. In addition, it is sometimes difficult for an operation to tell in advance what kind of lock it will need; for example, many types have operations that normally "write" an object, but sometimes only "read" (e.g., a *withdraw* operation on a bank account modifies the account if the balance is high enough, but otherwise only reads the current balance). This indicates that an automatic locking mechanism might be difficult to use. Also, the recovery mechanisms in [SCHW84] are complex, and place serious restrictions on the concurrency among activities that can be achieved.

Work at Newcastle on recovery blocks [ALS78] [ANDE79] [VERH76] investigated recovery techniques for building user-defined data types. While concurrency was not considered, alternative program structures were explored. The *inclusive recovery* scheme in [ALS78] is similar to our implicit approach, while their *disjoint recovery* scheme is similar to our explicit approach. The authors of [ALS78] [ANDE79] note that the inclusive scheme provides limited control over recovery and can be less efficient than the disjoint scheme, but that implementations in the disjoint scheme can be more complex than those in the inclusive scheme. These conclusions are similar to our conclusions about the implicit and explicit approaches for implementing atomic types.

Allchin [AM83] [ALLC83] has also investigated "atomic objects." He has focused on the implicit approach, attempting to make the system do as much of the work as possible. As we have discussed above, this approach provides limited expressive power. One interesting mechanism that he proposes is a queuing mechanism for an implicit scheme: He associates one queue with each object. An activity can wait on a queue, and will be awakened when another activity that used the same object becomes visible to it. The mechanism proposed by Allchin works well for simple examples, such as the semiqueue implementation above, that only need a single queue. His mechanism does not work, however, for more complicated examples that need multiple queues or a dynamically varying number of queues.

6. CONCLUSIONS

In this paper we have discussed techniques for implementing atomic types and linguistic constructs that support those techniques. We presented an example that illustrates a number of limitations of the techniques supported by Argus, and presented an alternative approach that avoids those limitations. The alternative approach suggested here, however, is still complicated; it is not clear whether an approach can be found that avoids some of this complexity, or whether the complexity is inherent in the problem domain.

There are a number of interesting areas for further work. Both the constructs in Argus and the alternatives suggested here are aimed at dynamic atomic types. We have not yet studied the requirements of other kinds of types, such as those that use a timestamp-based protocol (e.g., see [REED78]) to synchronize concurrent activities. We expect, however, that the explicit approach suggested here will be easier to extend to other kinds of concurrency control protocols than the implicit approach supported by Argus.

To ensure that the effects of committed activities survive crashes, the objects shared by activities must also be *resilient*, i.e., it must be extremely unlikely that data will be lost in a crash. The support for resilience in Argus, based on stable storage and extensions to the **mutex** type, is described in [WL85]; similar extensions work for the alternative constructs suggested in this paper. It would be useful to study alternative ways of implementing resilience, both different ways of using stable storage and ways of providing resilience through replication instead of stable storage.

Another problem that needs to be studied is deadlocks. Activities that share atomic objects can deadlock waiting for access to the objects. A way of detecting deadlocks is needed. Numerous algorithms exist for detecting deadlocks in a distributed system. However, since the synchronization constraints for user-defined atomic types are implemented by user code, the system cannot always know what an activity is waiting for. Thus, the user program must be involved in deadlock detection at least to some extent.

Neither the linguistic support in Argus, nor the alternative we suggested in this paper, supports optimistic implementations. It appears relatively easy to do so using an explicit approach, simply by allowing programmers to supply an explicit "pre-commit" operation for objects to vote on whether to allow an activity to commit. It appears very difficult to extend an implicit approach to permit optimistic implementations.

7. ACKNOWLEDGEMENTS

The author gratefully acknowledges the many perceptive comments and suggestions contributed by his thesis committee, Professors Barbara Liskov, John Guttag, and Nancy Lynch. The members of the Argus design group, especially Maurice Herlihy, Gary Leavens, and Sheng-Yang Chiu, also made numerous helpful suggestions.

212

REFERENCES

[ALLC83] [ALS78] [AM83] [ANDE79] [BGL81] [BR81] [DAVI73]
[DAVI78] [EGLT76] [KORT81] [LISK81] [LISK82] [LISK83] [LS83]
[LYNC82] [LZ74] [MOSS81] [REED78] [SCHW84] [SS84] [VERH76]
[WEIH83] [WEIH84] [WL85]

Chapter 14

Building Flexible Multilevel Transactions in a Distributed Persistent Environment

G. Lawrence Krablin
Glasgow University

University of Pennsylvania
Unisys Corporation

1. INTRODUCTION

As attempts [AB85] are made to integrate general purpose programming languages with the long term facilities of data definition and storage provided by database systems, the old problems of managing concurrent access and update become even more difficult. This paper will discuss these problems in the context of an integrated persistent environment and develop a flexible programming model to deal with them.

The integrated environment is based on a system supporting one or more programming languages. Such a language provides for the definition and manipulation of objects distinguished by type and value (including procedures). It supports data and procedural abstraction, including conventional scope rules restricting the visibility of named objects in the structure of the procedural abstractions. It is unconventional in decoupling the lifetime (or "persistence") of objects, with their values, from those scope rules. That is, the persistence of an object is not bound by the lifetime of an activation of the procedure or program which creates (declares) it. In contrast to the use of an external filing system, the persistent system retains intact the type and value of the object, as well as preserving the integrity of its relationships with other objects.

Another characteristic of the programming environment is the availability of cheap local concurrent processes. Processes are supported directly within the languages, as a natural extension of procedure invocation. Shared access to objects is constrained only by the corresponding scope rules. Communication among otherwise unrelated processes (separate "programs" executed concurrently) may take place through shared access to persistent data objects, or, in a more distributed model, directly by message passing, remote procedure call, rendezvous, etc.

In contrast, the familiar DBMS is not integrated with the programming environment. It constrains, through the schema, the types and values of persistent data objects, and restricts the relationships that may be preserved. The interface between the objects in the persistent store and programs with their transitory objects is through a separate Data Manipulation Language (DML) which translates the types and values of the persistent objects and does not allow relationships among persistent and transitory objects to be represented directly at all.

In the DBMS environment, management of concurrent processes with shared access to the database is done by a (more or less) centralized monitor, through transactions. The monitor must be able to recognize the boundaries of the transaction, through appropriate DML constructs, and, during a mix of concurrent transactions, identify any given action on the database with a particular transaction. This is typically done by associating transactions with whatever model of procedure, process, job, or program is provided by the underlying operating system. The central monitor approach seems to be inadequate for several reasons.

1. The DBMS typically restricts the abstraction facilities which may be applied to the persistent objects. It may not be obvious how this is related to management of concurrent transactions, but we will see that different levels of abstraction and implementation may have different requirements for concurrent interaction.

2. Standard database systems, and even some more flexible integrated systems [LISK84] [PS85] have a rigid and narrow notion of transaction. As mentioned above, it is generally closely bound to the model of procedure or process presented by the language and/or operating system. This makes it difficult to develop programs where concurrent processes may cooperate rather than compete, or do both simultaneously at different levels of abstraction.

3. Total concurrency control (serialization) is exercised by the DBMS on all data shared explicitly (in the database) by concurrent processes, without regard to the roles being played by various objects in the interaction. When, as in the separate DBMS model, the persistent data is kept totally separate from the transient objects in the program, this problem is not severe. The programming language/system itself provides tools (such as semaphores [DIJK68] and monitors [HOAR74] [BRIN75]) to regulate shared usage of the transient data. In an integrated system, however, there is no neat line drawn (nor can one be drawn in general) to distinguish persistent from transient data. To draw such a line would require different semantics for objects which are otherwise alike in type and value. On the other hand, to apply strict concurrency control (via transactions) to all shared data severely limits the degree of cooperation available and unduly restricts potential concurrency.

4. A useful and common property of database transactions is that their effects are undone or rolled back if the transaction fails (whatever that means and however it is detected). As with concurrency control, either all the effects must be rolled back or there must be some mechanism to decide what should be rolled back. The conventional DBMS simply rolls back the state of the database and ignores the program except to signal the failure. Again, in the integrated persistent environment, this line can't be drawn by the system. Rolling back everything, on the other hand, defeats the purpose of audit trails, logs, etc.

5. Structured programming techniques call for breaking a task, such as a transaction, into smaller, more managable units. A logical extension of the procedural abstraction mechanism is to provide these smaller units with the same properties as those expected for the whole transaction, such as rollback on failure or proper control of competing concurrent access. The introduction of sub-transactions makes different demands on these properties, in order to support family trees of transactions. The conventional DBMS paradigm makes no provision for sub-transactions.

6. The strict concurrency control, auditing, and rollback mechanisms of a conventional DBMS make no provision for transactions of widely varying duration. Transactions extending over a long period of time (in CAD work, for example) may eventually cause complete paralysis of other processes using the database. The consequences of forced rollback (on deadlock, for example) would be quite unpleasant. However, if transactions are to be a preferred abstraction for dealing with long lived shared data, this situation must be adequately dealt with.

7. The transaction implementation commonly provided by a DBMS does not nicely extend into distributed systems and systems where independent transactions communicate with one another. A framework is necessary where update and commit can be handled properly, and consistent action taken when a transaction fails, having caused the update of data on another host in a network, or, through communication, affected the course of another transaction.

It is interesting to note that many of the points mentioned above arise implicitly in the implementation of a DBMS. Control of concurrent access by unrelated processes requires finer grain control on concurrent use of data objects used in the implementation. Rollback is naturally more flexible to allow logging of failures, and to provide for crash recovery. Preparation for crash recovery also brings in the use of "two-phase commit"[GRAY78], a protocol usually associated with distributed systems. If these properties are required for the implementation of a DBMS, it follows that they must also be provided in a language designed to support the construction of even more flexible systems.

The point is that the conventional model for shared databases does not allow for concurrent cooperative activities to take place at different levels of abstraction and implementation within a single program or a group of programs. The rest of this paper will describe and demonstrate a straightforward and flexible paradigm for building transaction systems with the properties needed for specific situations.

2. PRIMITIVE SUPPORT

The context of the examples and discussion in the remainder of this paper is that of PS-algol and an extended version of it, CPS-algol. These language systems support an environment where objects may persist as long as potential references to them remain. The following sections sketch salient features of the languages.

2.1. PS-algol

For the purposes of this paper, there are two important distinguishing characteristics to PS-algol [PS85]: persistence and first class procedures. Persistence is treated as an orthogonal attribute of any data item, in the sense that such an item remains in the heap (or the persistent store) as long as there are references to it. An item may be an object in the heap, referred to by explicit pointers appearing in other data objects or in procedure frames, or it may be a simple variable in a procedure frame itself. In the latter case, items in a frame are accessible if the frame is part of the static or dynamic chain of an active process or is part of the static environment of an accessible procedure.

The PS-algol system has a number of distinguished nodes, called databases, which are permanently reachable (in the permanent store on disk) by name, rather than by explicit pointers. The topmost structure in a database, a table, is a symbolic association of strings or integers to pointers. Objects in the permanent store are brought into the heap as they are referenced by a program. They are written back, along with any objects newly reachable from the persistent store, when the program performs a commit. If the program aborts, it is abandoned completely, along with the current contents of the heap.

Procedures in PS-algol are fully first class objects [ABCCM84]. They may be assigned to variables, returned as functional results, and stored in data structures. A procedure consists of its code and a pointer to the frame corresponding to its immediate global environment. Frames are linked statically by pointers so that all components of the procedure are reachable. With this approach, a procedure reachable from a database may be placed in the permanent store (with its closure), to be invoked later by other programs.

First class procedures are the basis for a useful packaging technique for abstract data objects [AM84], which will be employed in later examples. A 'make' procedure is defined which contains the data and procedures implementing the abstract data item. It returns as its result a pointer to a data structure containing only the exported procedures (operations). The data remains in the frame of the make procedure, accessible only to the exported operations.

2.2. CPS-algol

CPS-algol [KRAB85] is a version of PS-algol augmented with constructs to support and manage concurrent processes. It was designed to provide experimental support for the work described in this paper. The two main features of CPS-algol are multiple independent processes running concurrently, and special data objects providing primitive concurrency control.

Any procedure of type void (returning no result) may be supplied with appropriate arguments and started as an independent process. It will continue to run until the procedure on which it is based completes, a suicide primitive is executed, or a program error causes a crash of the process. The addressing environment is exactly as if the procedure had simply been called.

A procedure requiring no arguments and returning no result may be called a 'proto-process'. Any procedure which does take arguments may be packaged with a set of actual arguments as a proto-process by a procedure which captures the procedure and its arguments, and returns a simple procedure calling the captured procedure with its arguments. A proto-process is a convenient object to use for concurrency, as it may be passed around without type problems before being started as a process by an operation of an abstract module.

When a process is started, a pointer to an object ("process.strc") with two visible fields is returned. One indicates the status of the process: alive, finished, suicided, or crashed. The other is a pointer which may be supplied as part of the start primitive. This utility pointer may be used to attach any sort of object to the process. These fields may be accessed through the pointer returned from start, or the pointer provided by current.process, which always refers to the process within which it is invoked.

A final aspect of CPS-algol processes is the epilog procedure. A procedure (no arguments or result) may be supplied when the process is started. It will be called by the system when the process terminates. It has no special privileges, but it is run as part of the original process, so it can refer to the status and utility fields asociated with the process, using current.process. The final status is set before the epilog procedure is called.

The special data items in CPS-algol are called resource objects and are defined with fields and referenced through pointers just as are ordinary PS-algol structure objects. The difference is that the fields in a resource object may only be used within the dynamic scope of an atomic clause (a critical region) defined for that object. The atomic clause guarantees exclusive access to the object for that process. An await clause allows critical regions to be conditional.

It is worth noting that a process and the data structures associated with it may be made persistent if a commit is done while the process is running and its process.strc is reachable from the database. If a later access to the database "touches" the process, it will be revived (if it isn't already running) and will continue to run at the point when the commit happened.

3. TRANSACTIONS

A database, as an abstraction representing some aspects of some enterprise, has a structure defining the nature of the data it contains and the relationships among various contained data objects. The structure and other constraints implied by the enterprise (banking regulations, for example) may be considered to be a consistency criterion for the database. For applications built around database systems, transactions [DAVI78] [DATE83a], as basic units of work,form major building blocks. An important requirement of a transaction is that it maintain the consistency of the database, even though in the midst of the sequence of activities making up the transaction, the database may go through intermediate states where it fails to meet its requirement of a transaction is that it maintain the consistency of the database, even though in the midst of the sequence of activities making up the transaction, the database may go through intermediate states where it fails to meet its consistency criterion.

3.1. Transaction Properties

There are a number of properties demanded of transactions in order to make them useful for building complex applications systems. The first such property is identity. In both human affairs and computer database systems, a transaction may comprise a number of separate activities. In an environment where a number of things may be going on concurrently (on behalf of various unrelated transactions), it is important to be able to identify any given action with the correct transaction. This is implicit in all work on transactions, but is not often explicitly noted in the literature.

In keeping with the role of a transaction as a consistency preserving unit of work, an important characteristic is that it be atomic. That is, a transaction must always appear as if it has not yet (or never) started or has successfully completed. This characteristic is described in terms of two properties: reversibility and serializability.

In a transaction made up of many different activities, there are numerous ways for things to go wrong, making successful completion impossible. If the transaction is to be atomic, there must be some mechanism to prevent any intermediate effects (rendering the database inconsistent, either internally or in terms of the abstraction of the real world) from being made permanent. A reversible transaction may delay all effects to the end or provide for such effects to be reversed (backed out) in case of failure.

In any environment supporting a serial stream of transactions, each transactions sees and acts on the database as it is left by the preceding transaction. Since transactions are (intended to be) consistency preserving, a transaction is always assumed to begin with the database in a consistent state. When the transactions are acting concurrently, however, they start while others are in the middle of processing, making consistency preserving atomicity problematical. Various disciplines [EGLT76] [KR81] [REED78] [BSW79] [BG80] [BG83] have been developed and studied to guarantee serializability - that is, that the interactions of a concurrent mix of transactions will be restricted such that the effect of the total mix is identical to some serial ordering of the transactions, thus allowing the inductive guarantee of consistency.

Finally, transactions do not operate solely in the environment of the database, but communicate results to external agencies in the form of printed reports (or checks!), messages to other computer systems or humans, cash dispensed, etc, all communications which may be difficult or impossible to reverse. It may be possible [DATE83a] to delay these communications until the transaction is complete, but there also needs to be some assurance that it won't later be undone, either by "normal" activities such as rollback, or untoward events like disk crashes. Transactions are robust (or recoverable) to the extent that, having signalled successful completion, they stay completed (in terms of their effects on the database), in spite of at least some kinds of catastrophes.

3.2. Primitive Transactions

In CPS-algol (and other languages supporting similar constructs), a critical region may be viewed as a primitive transaction. The "database" is the resource on which the critical region is defined. The definition of critical region implies serializability, as it is impossible for two critical regions (on the same resource) to be active simultaneously.

The only ways for a critical region to fail are for the process to suicide or crash, or for the underlying system to crash. In neither of these cases is any recovery undertaken by the CPS-algol system, and the resource may be left permanently inaccessible (locked). The system makes no effort to identify the critical region as a transaction or to keep track of its activities, so the primitive transaction is neither reversible nor robust. However, neither programming failures (as opposed to system crashes) nor explicit aborts are expected at the primitive level of abstraction where critical regions are used. Communication of results by the primitive transaction is generally through the resource itself or data objects local to the process containing the critical region. A system crash would certainly obliterate the latter, and any recovery would restore the resource to a state before the critical region (assuming that the critical regions are appropriately constructed).

4. BUILDING TRANSACTIONS AND OBJECTS

In the following sections, the primitive facilities provided by CPS-algol will be used to build objects and transactions with useful properties. These properties will include atomicity and reversibility of transactions with respect to shared objects. Different potential relationships among transactions will be explored.

4.1. Simple Objects and Independent Transactions

The example given here provides two phase locking [EGLT76] for serialization, and the effects of a transaction on shared objects are fully reversible. In this simple model, a process serves as a transaction, and neither communications among transactions nor nested sub-transactions are provided for. Also, no provision is made in this example for recovery from system crashes, only local transaction failures.

A transaction is represented as an abstract data object attached to a process through its utility pointer. The epilog procedure for the process is used to ensure that if the transaction isn't committed or aborted explicitly, it will be aborted. The data encapsulated by the transaction object is just the transaction status, a lock to guarantee atomicity of transaction operations, and a pair of action lists, one for commit, and the other for abort. The operations are commit, abort, and log, which adds a pair of actions (one for commit, the other for abort) to the list of things to be done when the transaction terminates. The existence of a list (of pointers) data type is assumed, with distinguished value nil.list, operations cons, hd, tl, and predicate null.

A transaction is represented by a PS-algol data structure:

```
structure transaction( proc() trans.commit,trans.abort ;
                       proc( proc(pntr),proc(pntr) ) trans.log )
```

and a procedure for creating new transactions:

```
let mk.trans = proc( -> pntr )
begin
     resource locker
     let lock = locker
     structure encap.act( proc( pntr ) action )

     let active = 0
     let committed = 1
     let aborted = 2

     let status := active
     let commit.actions := nil.list
     let abort.actions := nil.list
     let id := nil

     let do.commit = proc()
     atomic lock do
     if status = active do
     begin
          while ~null( commit.actions ) do
          begin
               hd( commit.actions )( action )( id )
               commit.actions := tl( commit.actions )
          end
          status := committed
     end

     let do.abort = proc()
     atomic lock do
     if status = active do
     begin
          while ~null( abort.actions ) do
          begin
               hd( abort.actions )( action )( id )
               abort.actions := tl( abort.actions )
          end
          status := aborted
     end

     let do.log = proc( proc( pntr ) c.act,a.act )
     atomic lock do
     if status = active do
     begin
          commit.actions := cons( encap.act( c.act ),commit.actions )
          abort.actions := cons( encap.act( a.act ),abort.actions )
     end
```

```
    id := transaction( do.commit,do.abort,do.log )
    id
end
```

Attached to each process of interest will be the epilog procedure:

```
let trans.epi = proc()
begin
    let trans.id = current.process( process.util )
    if current.process( process.status ) isnt process.alive then
    trans.id( trans.abort )() else
    suicide                    ! the epilog was called incorrectly
end
```

Under the assumption that a transaction is a process, the following procedure will start one:

```
let start.trans = proc( proc() proto.process -> pntr )
begin
    let tran = mk.trans()
    start proto.process() with tran epilog trans.epi
end
```

and the process can commit or abort by calling:

```
current.process( process.util,trans.commit )()
```

or

```
current.process( process.util,trans.abort )()
```

or can abort implicitly by terminating without doing either, in which case the epilog forces the abort. It would be straightforward to change the epilog procedure to commit by default on normal termination instead of aborting it.

The framework described above does nothing about issues of concurrency control (except to protect itself) nor about assuring that the effects of a transaction will be cleaned up if it fails. These responsibilities are left to the objects affected by them. All shared objects in this example are treated as instantiations of abstract data types, with some state and a set of explicit operations to manipulate that state. Issues of atomicity and reversibility are managed by the objects, using the tools provided by transactions. In the sample object implementation below, a set (of pointers) data type is assumed, with distinguished value empty.set, operations add.element, remove.element, and predicates member, empty, and singleton (singleton (s, e) <=> s is the singleton set containing e).

The object, which is just a simple integer variable, is represented by a structure containing its operations:

```
structure int.object( proc ( -> int ) read.int ;
                       proc( int ) write.int )
```

and a procedure for creating new instances:

```
let mk.int = proc( -> pntr )
begin
     resource locker
     let lock = locker
     let value := 0
     let perm.value := 0
     let writer := nil
     let readers := empty.set

     let read.finish = proc( pntr trans.id )
     atomic lock do
     readers := remove.element( trans.id, readers )

     let write.commit = proc( pntr trans.id )
     atomic lock do
     begin
         writer := nil
         readers := nil.list
     end

     let write.abort = proc( pntr trans.id )
     atomic lock do
     begin
         writer := nil
         readers := nil.list
         value := perm.value
     end

     let do.read = proc( -> int )
     begin
         let trans.id = current.process( process.util )
          atomic lock do
          begin
              await member( trans.id, readers ) or writer = nil
              if ~member( trans.id, readers ) do
              begin
                  readers := add.element( trans.id, readers )
                  trans.id( trans.log )( read.finish, read.finish )
              end
              value
         end
     end

     let do.write = proc( int new.value )
     begin
         let trans.id = current.process( process.util )
         atomic lock do
         begin
              await writer = trans.id or
                    ( writer = nil and
                    ( empty( readers ) or
                      singleton( readers, trans.id ) ) )
              if writer ~= trans.id do
              begin
                  writer := trans.id
                  trans.id( trans.log )( write.commit, write.abort )
```

222

```
                readers := add.element( trans.id, readers )
                perm.value := value
            end
            value := new.value
        end
    end

    int.object( do.read, do.write )
end
```

It should be evident that a process started using start.trans will be atomic with respect to objects created by mk.int. The serialization technique used allows multiple concurrent readers, but write permission is exclusive. The operations on int.object could be extended to require explicit lock requests instead of the implicit locking provided. Because unlocking is put off until the end of the transaction, any number of objects using similar techniques will collectively implement two phase locking.

For a simple object like this one, controlling concurrent access and restoring the value when a transaction aborts are quite straightforward. For a composite object containing pointers to other objects, etc., the question arises of defining the boundaries of the object for purposes of concurrency control and rollback. These boundaries are properly attributes of the abstractions represented by the objects, not the usage of them by transactions. As such, this model for objects allows complete flexibility in such matters. It is likely for large, complex objects, and particularly for large primitive objects, that simply making a copy (to restore on abort) will not be a viable method. Objects may be implemented to use change lists, paging, and other techniques to achieve the same purpose. In a language providing primitive support for bulky or complex objects (relations, graphics images, etc.), some such methods may be built-in.

It may appear that there is a great deal of code which must be written to implement transactions and objects this way. To some extent, this is the price of flexibility, although there is a great deal of code also required to implement a general purpose DBMS. The amount of code could be significantly reduced if it was possible to write generic routines and use 'class hierarchy' construction methods [BDMN73] [GR83] to produce canned routines. Also, as suggested above, some disciplines could be provided as primitives by the language.

Any number of transactions may be started in parallel by successive calls on start.trans. They will compete for the use of objects (and possibly deadlock - see later section). If two transactions are to be explicitly ordered in time, the start of the second can be delayed until the first has completed. Transactions could be extended to provide this service directly, or the invoking process could wait for a completion status on the first transaction process.

Another alternative is to allow a single process to identify itself with a series of transactions in a sequential manner. This is easily done by adding an operation to attach a transaction to a specific process, with appropriate safeguards to avoid dangling transactions or two processes with the same transaction. Even this latter situation could be allowed by refining the epilog, commit, and abort procedures.

A transaction in progress may also start another asynchronous and independent transaction by calling start.trans at any time. The second transaction may be properly nested in time ('nested top level transaction' [LISK84]) by delaying the outer one until completion of the inner one. Again, the transaction abstraction could easily be extended to manage this directly, within the same process.

In all of these situations, the transactions are independent of one another in the sense that they compete for exclusive usage of the shared objects (like int.object in the example) and cannot communicate through these objects except sequentially. Stepping down a level, it can be seen that the individual critical regions in the implementations of transactions and objects have exactly the same properties with respect to the data items used to construct the objects. Each critical region is a separate top level nested transaction within that abstraction. These independent transactions implement the higher level transactions and objects by carrying the high level state (committed and in-progress versions) as the committed value of the low level state. Low level transactions, lacking important properties (such as recovery from failure) have thus been used to implement a useful, recoverable abstraction. This has been done entirely within the type system of the language.

This approach in CPS-algol is somewhat in contrast to that presented for ARGUS in [WEIH85]. ARGUS already has a well developed transaction structure, which is further elaborated to provide a similar degree of flexibility.

4.2. Dependent Sub-Transactions

The relationships among transactions as described in the previous section fell into two categories. In one case, they compete for the same objects. In the second, they are defined at different levels of abstraction, and the lower level implements the higher level. In either case, one might be a sub-transaction of the other in terms of timing, but in terms of their effects on controlled objects and the global visibility of those effects, they are independent.

Next to be considered are transactions which are nested in time, but not fully independent [MOSS81]. During the progress of a transaction, it operates in an environment of data objects. These are the objects which are entirely local to the transaction, and those more global, shared, objects it modifies or refers to. A dependent transaction operates within the environment of the parent. This has two consequences. The dependent transaction is affected by changes made previously to objects by the parent, and, to the extent that objects used by the child are atomic (managed for concurrent access, as was int.object above), the effects of the child are visible to the parent when the child commits, but not to the outer world (of transactions competing with the parent) until the parent commits. From the outside, the dependent transaction is an indistinguishable component of the parent. From within the parent, it is just another atomic action used to implement the parent, but possibly with the properties of reversal on failure and serialization among a number of concurrent (sibling) dependent transactions, making it a useful building block.

In order to build dependent transactions, the implementation of both objects and transactions must be extended. A transaction must now have additional operations to create a sub-transaction and to report on the ancestral relationships among transactions:

```
structure transaction( proc() trans.commit,trans.abort ;
                        proc( proc(pntr),proc(pntr) ) trans.log ;
                        proc( -> pntr ) mk.sub.trans ;
                        pntr trans.parent ;
                        proc( pntr -> bool ) is.ancestor
                            ! is this an ancestor of mine? )
```

The operations provided by objects such as int.object do not change, but the internal representation and management of concurrency must be modified to handle the requirements of dependent transactions. A series of previous values must be maintained to allow a single sub-transaction to abort. The sub.commit procedure will pass ownership of the object up one level to the parent. In case of abort, the object will be left in the state it was in before that dependent transaction started. The procedures below replace or augment those in mk.int:

```
structure old.int( int prev.value ; pntr prev.trans )

let old.values := nil.list

let read.sub.commit = proc( pntr trans.id )
begin
     let parent = trans.id( trans.parent )
     atomic lock do
     begin
          readers := remove.element( trans.id,readers )
          if ~member( parent,readers ) do
          begin
               readers := add.element( parent,readers )
               if parent( trans.parent ) = nil then
               parent( trans.log )( read.finish,read.finish ) else
               parent( trans.log )( read.sub.commit,read.finish )
          end
     end
end

let do.read = proc( -> int )
begin
     let trans.id = current.process( process.util )
     atomic lock do
     begin
          await member( trans.id,readers ) or
               writer = nil or
               trans.id( is.ancestor )( writer )
          if ~member( trans.id,readers ) do
          begin
               readers := add.element( trans.id,readers )
               if trans.id( trans.parent ) = nil then
               trans.id( trans.log )( read.finish,read.finish ) else
               trans.id( trans.log )( read.sub.commit,read.finish )
          end
          value
     end
end

let write.sub.commit := proc( pntr trans.id )
begin
     let parent = trans.id( trans.parent )
     atomic lock do
     begin
          let old = hd( old.values )
          if old( prev.trans ) = parent then
          old.values := tl( old.values ) else
          begin
               if parent( trans.parent ) = nil then
               parent( trans.log )( write.commit,write.abort ) else
               parent( trans.log )( write.sub.commit,write.abort )
```

```
                    readers := add.element( parent,readers )
               end
          writer := parent
       end
end

let write.abort = proc( pntr trans.id )
atomic lock do
begin
     if null( old.values ) then
     begin
          value := perm.value
          writer := nil
     end else
     begin
          let old = hd( old.values )
          old.values := tl( old.values )
          writer := old( prev.trans )
          value := old( prev.value )
     end
     readers := remove.element( trans.id,readers )
end

let do.write = proc( int new.value )
begin
     let trans.id = current.process( process.util )

     let all.ancestors = proc( -> bool )
     begin
          let list := readers
          while ~null( list ) and
               ( hd( list ) = trans.id or
               trans.id( is.ancestor )( hd( list ) ) ) do
          list := tl( list )
          null( list )
     end

     atomic lock do
     begin
          await writer = trans.id or
               trans.id( is.ancestor )( writer ) or
               ( writer = nil and
               ( empty( readers ) or all.ancestors() ) )
          if writer ~= trans.id do
          begin
               if writer ~= nil then
               old.values := cons( old.int( value,writer ),old.values)
               else perm.value := value

               if trans.id( trans.parent ) = nil then
               trans.id( trans.log )( write.commit,write.abort ) else
               trans.id( trans.log )( write.sub.commit,write.abort )

               writer := trans.id
               readers := add.element( trans.id,readers )
          end

          value := new.value
     end
end
```

Nearly all the changes to support dependent transactions have been made to the objects to implement the nested locking discipline. This is appropriate, as, in this model, the management of the concurrent environment is distributed among the objects forming the environment.

The algorithms in this example will manage various combinations of concurrent transactions partially composed of concurrent sub-transactions, etc. One area of potential trouble is a transaction running concurrently with one of its own dependent sub-transactions. If they compete for the use of an object, the child will override the parent's write lock (as it must to avoid deadlock), but a parental read lock will continue to be valid even when the child has a write lock. This will violate the atomicity of the child. The locking routines could be refined to prevent this specific case, but the general situation seems to be pathological and is best avoided by preventing such competition altogether. (In the paradigm of a sub-transaction being a building block for the parent, there is no sense of the parent having a separate identity). Checking that a transaction has no active dependents before granting object access is one way of accomplishing this.

On the other hand, a parent and a dependent child might not compete at all for objects. In this case, the dependency aspect arises only when they finish. When the child commits, its objects are passed on to the parent, with final commit being delayed until the top level transaction commits. A parent cannot commit until all of its dependents have completed. A count of active dependents may be maintained by giving each dependent (when it is created) a post-action to decrement this count. A parent which may only commit if all (or some particular) dependents also successfully commit, may log a different action to each such dependent for abort, to force the eventual abort of the parent.

4.3. Using Transaction and Object Properties

In the previous sections, a basic paradigm for the relationship between transactions and objects has been described. The examples given demonstrate some combinations of transactions and objects with properties similar to those of standard database systems. Other combinations could be used; for example, if a dependent sub-transaction used an object like the original int.object, the commit of the dependent would be final (with respect to that object), without regard to the final disposition of the parent. This seems appropriate for, among other things, logging and journalling.

A different kind of object could reach further up the family tree of a transaction, to plant post-actions at higher levels. Since these actions will be performed in the context of the object state as it is at that time (probably different from the state when the actions were logged), considerable flexibility is gained to recover from faults and exceptions relating to the objects, as well as to rollback an aborted transaction. As shown in a later section, communications among objects and independent transactions can be managed in the same way.

The object presented in the example above used locking to implement serialization. Other disciplines have been developed to serve the same purpose. Multiple versions [BG83] and timestamping [REED78] [BG80] could be applied in the same framework, although more information would have to be maintained with the transaction itself. Optimistic methods [KR81] would be more difficult to implement in this decentralized environment, especially since records must be kept indefinitely of the activities of transactions already successfully completed. A central register of transactions would probably be required, but isn't necessary at all with the scheme presented here.

4.4. A More Elaborate Example

The next example is borrowed from the ARGUS project [WL85] [WEIH85]. A semiqueue is a data structure with properties somewhat like those of a queue. Items may be enqueued and later dequeued. However, a semiqueue does not require that items be dequeued in the same order as they were enqueued, nor in any other particular order. This nondeterministic choice of object for dequeuing is used to allow greater concurrency than would be possible with a strict FIFO discipline, since there is no need to wait for the transaction which placed the first item in the queue to commit.

The implementation given below of a semiqueue uses a double linked list, with enter and delink procedures. Concurrent transactions may freely call the two exported operations, enqueue and dequeue, with the critical regions protecting the head, tail, and count variables used to represent the list. Individual entries do not themselves need additional protection (for changes to the status and owner fields), as there is always implicit ownership conveyed with the commit and abort postactions.

An enqueue operation can always proceed immediately, as all it requires is internal consistency of the semiqueue. A dequeue operation must wait for the availability (expressed in the procedure got.one) of a suitable entry. This version of a semiqueue is designed to be used with dependent sub-transactions. A newly inserted entry may be removed by the same transaction or any descendent, or if the inserting transaction commits, an ancestor. The various commit and abort postactions properly promote ownership or restore the state of the entry. If an entry is deleted by the same transaction that inserted it, or by an ancestor of that transaction, no post action is necessary.

The whole structure is self-contained, depending only on the simple properties of the transaction abstraction. All the serialization and reversibility qualities are directly expressed in terms of the state of any given entry, the requirements for dequeuing, and the postactions referring to that entry.

Note that the procedures q.add.commit, q.add.abort, q.delete.commit, q.delete.abort, q.del.ins.commit, and q.del.ins.abort are not themselves the postactions, but take the entry and bind it to the actual postaction. The postaction is then returned and passed on to trans.log.

```
structure semi.queue( proc(pntr) enqueue ; proc( -> pntr ) dequeue )

let mk.semi.queue = proc( -> pntr )
begin
      resource locker
      let lock = locker
      let count := 0
      let head := nil
      let tail := nil

      structure s.q.item( pntr f.link,b.link,item,owner; int status )

      ! values for status
      let inserted = 0
      let committed = 1
```

```
let deleted = 2

let delink = proc( pntr entry )
begin
    if count = 1 then
    begin
        head := nil
        tail := nil
    end else
    if tail = entry then
    begin
        tail := entry( b.link )
        tail( f.link ) := nil
    end else
    if head = entry then
    begin
        head := entry( f.link )
        head( b.link ) := nil
    end else
    begin
        entry( f.link,b.link ) := entry( b.link )
        entry( b.link,f.link ) := entry( f.link )
    end
    count := count - 1
end

let enter = proc( pntr entry )
begin
    if count = 0 then head := entry else tail( f.link ) := entry
    entry( b.link ) := tail
    entry( f.link ) := nil
    tail := entry
    count := count + 1
end

let q.add.commit = proc( pntr entry -> proc( pntr ) )
proc( pntr trans.id )
if entry( status ) = inserted do
begin
    let parent = trans.id( trans.parent )
    if parent = nil then entry( status ) := committed else
    begin
        entry( owner ) := parent
        parent( trans.log )( q.add.commit( entry ),
                             q.add.abort( entry ) )
    end
end

let q.add.abort = proc( pntr entry -> proc( pntr ) )
proc( pntr trans.id )
if entry( status ) = inserted do
atomic lock do
delink( entry )

let q.delete.commit = proc( pntr entry -> proc( pntr ) )
proc( pntr trans.id )
```

```
begin
    let parent = trans.id( trans.parent )
    if parent = nil then atomic lock do delink( entry ) else
    parent( trans.log )( q.delete.commit( entry ),
                         q.delete.abort( entry ) )
end

let q.delete.abort = proc( pntr entry -> proc( pntr ) )
proc( pntr trans.id )
entry( status ) := committed

let q.del.ins.commit = proc( pntr entry -> proc( pntr ) )
proc( pntr trans.id )
begin
    let parent = trans.id( trans.parent )
    if entry( owner ) = parent then
    atomic lock do delink( entry ) else
    parent( trans.log )( q.del.ins.commit( entry ),
                         q.del.ins.abort( entry ) )
end

let q.del.ins.abort = proc( pntr entry -> proc( pntr ) )
proc( pntr trans.id )
status( entry ) := inserted

let enq = proc( pntr data )
begin
    let trans.id = current.process( process.util )
    atomic lock do
    begin
        let entry = s.q.item( nil,nil,data,trans.id,inserted )
        enter( entry )
        trans.id( trans.log )( q.add.commit( entry ),
                               q.add.abort( entry ) )
    end
end

let deq = proc( -> pntr )
begin
    let trans.id = current.process( process.util )
    atomic lock do
    begin
        let temp := nil

        let got.one = proc( -> bool )
        begin
            let t := count
            let done := false
            temp := head
            while t > 0 and ~done do
            if temp( status ) = committed or
               (temp( status ) = inserted and
               (temp( owner ) = trans.id or
                trans.id( is.ancestor )( temp( owner ) ) ) ) )
            then done := true else
            begin
```

```
                              t := t - 1
                              temp := temp( f.link )
                    end
                  done
        end

        await got.one
        if temp( status ) = committed then
        trans.id( trans.log )( q.delete.commit( temp ),
                               q.delete.abort( temp ) ) else
        if temp( owner ) = trans.id then delink( temp ) else
        trans.id( trans.log )( q.del.ins.commit( temp ),
                               q.del.ins.abort( temp ) )
        temp( status ) := deleted
        temp( item )
      end
    end

    semi.queue( enq,deq )
end
```

4.5. Stability and Crash Recovery

In the examples above, provision is made only for failures to the extent of failure or abort of a transaction. In particular, a failure occurring within one of the primitive critical regions, including during commit/abort processing, would potentially be fatal to the whole system. At the very least, some resources and transactions would be irretrievably lost (permanently locked or forgotten). This characteristic extends to failure in any lower level abstraction which is used to implement a higher level abstraction. Some kinds of failures, such as overflows and other data errors, may be avoided by careful programming, or handled using lower level or dependent transactions.

A system crash, however, is beyond the programmer's ability to prevent, as it involves a failure of some component below the level of abstraction provided by the programming language (or the level being used to implement the abstraction under consideration). Recovery from such a failure requires the use of a recording medium not corrupted by the particular failure to be recovered from. It is obvious that the ability to recover from a crash is not absolute, but only relative to the particular failure. The typical approach [LAMP81] involves successively more permanent storage (primary memory, disk, tape, etc.) and/or replication (which improves the probability of recovery from a particular kind of failure within the same level).

PS-algol treats its address space as a two level store, with the permanent image on disk and a cached copy in primary memory. In the existing management scheme for this system, transactions are committed by copying all reachable objects onto the disk containing the committed version of the data. In the transaction structure suggested above, commit (the PS-algol primitive) may be used to ensure the relative permanence of the current state without any regard to its relationship with ongoing (abstract) transactions and sub-transactions. As was the case with using primitive critical regions to implement more useful transactions, PS-algol commit, in conjunction with individually tailored recovery procedures, can be used to ensure that enough information is directly recoverable to manage the higher level recovery process. Note that, since processes may be persistent, commit at any point preserves the complete state of the system, making recovery trivial.

PS-algol commit is a bit of a bludgeon for this purpose, as it makes everything permanent, whether is is needed or not. A central permanent store handler could be set up to provide better coordination, but this centralizes what has thus far been a decentralized system. Instead, a primitive command to stabilize (e.g., copy to disk) a single object (but not objects referred to by contained pointers) might be proposed. It fits very well into the general structure of transactions and objects, as the bounds of an abstract object (possibly including other objects pointed to) are determined by the individual abstraction.

It must be noted that the decentralized system of objects and transactions described here results in a multitude of independent "resource managers" (one for every object!), which must be coordinated [DATE83] [GRAY78] to perform commit. In this context, the need for the two-phase commit protocol becomes apparent. This protocol demands of its participants that certain information (agreement to commit, for instance) be remembered in spite of potential crashes and ensuing recovery. This requirement holds for the single system transactions and objects presented here and especially for system distributed over multiple computers, all subject to independent failures.

4.6. Deadlock

There are two kinds of deadlock which may arise in systems built like the examples above. The first is the possibility, within the implementation of objects and transactions, that deadlock will occur with respect to dynamically nested critical regions at the primitive level. The examples, as given, are believed to be free of such deadlock, but it would be very easy to introduce it when adding such refinements as were suggested.

The more general problem is that of higher level concurrent transactions competing for the use of objects (the case discussed in the previous paragraph is a primitive example, of course). The organization of objects and transactions suggested in the examples gives no way of preventing or detecting deadlock. The most obvious reason for this is that the structure described is a distributed one, while deadlock is a global property of the system.

It may be possible, within a given abstraction, to avoid deadlock by conventional means: ordering resources, pre-allocating resources, etc., but this seems infeasible in general. Given that deadlock situations at higher levels could be detected (by timeout, for example), there is already machinery in place to manage abort properly. An externally forced abort of a transaction is easy enough to implement, but must be accounted for wherever a process (acting for a transaction) may become blocked. This is a drawback of this distributed model.

5. OTHER ISSUES

It has been shown how this model can be used to deal with the classical database transaction issues of atomicity and reversibility with respect to shared objects. The same basic model can be applied to communication among transactions and objects, both locally and in a distributed system.

5.1. Communications

The simplest form of communication among transactions is the passing of arguments to a sub-transaction (dependent or independent) and receiving in return status and results. In the case of a dependent transaction, such communication may take place implicitly through the shared use of the parent's environment. For independent transactions, some other medium must be used. One of the refinements suggested in a previous section involved making the status of a transaction available for testing by another transaction. This test would, in effect, be an operation on the transaction object. It has already been noted that such operations are themselves independent transactions, at a lower level of abstraction. This suggests that communication be generally modelled in this way, the medium being an abstract object with send and receive (for example) operations.

Using independent transactions to implement communication relieves us of trying to specify at this level the relationships among the communicating transactions. In particular, the structure of the system imposes neither a coordinated commit process nor any sort of cascading abort. Instead, the programmer has the leeway to consider the nature of the communication and its effect on both transactions. Where cordination is required at termination, post-actions may be logged to the transactions, invoking an appropriate commit/abort protocol. In such cases, it may be useful to either redesign objects to provide a two phase commit, or to impose an artificial parent transaction on each of the partners, so that conditional committal may be done, verifying that the final commit will be successful (just a different version of two phase commit). In more loosely coupled circumstances, a failing transaction might inform communications partners of its own demise, again through post-actions logged to itself.

5.2. Distributed Systems

The model for objects and transactions that has been presented is already of a very distributed nature, but it relies heavily on procedure call and storage of first class procedures to acheive its ends. In a truly distributed system, such direct procedure calls may not be possible. Instead, a mechanism based entirely on communication must be devised.

In the first place, access by a transaction to an object on another node in the network will have to be accomplished (at some level) by a surrogate activity on the same node as the object. The surrogate can invoke operations, send a copy of the contents of the object, etc., for the original transaction. If the surrogate is implemented as a transaction, then it can be treated as a (remote) dependent sub-transaction sharing no objects with the parent. The problems of starting, communicating with, and properly terminating a remote sub-transaction may be dealt with through a lower level network facility.

This lower level network transaction facility can be thought of as providing transaction-to-transaction virtual circuits, including the service of starting a remote transaction. Sending and receiving messages, as in local communication, are treated as independent transactions. The virtual circuit abstract object straddles the two nodes involved and logs post-actions locally to the transactions involved. The post-actions are used to initiate and coordinate a distributed commit/abort protocol. At this point, local and non-local communication become very similar. Indeed, the network transaction abstraction would work just as well for purely local sub-transactions (not sharing data directly).

Local pointers referring to nonlocal objects are a potential problem with heap and pointer based systems like PS-algol. One approach is to base the entire network on a global address space and have the underlying implementation take care of communications needs. This hides the distributed nature of the system, at the cost of requiring built-in disciplines for coordinated commit, managing communications failures, etc.

More in keeping with the layered and distributed nature of the structures described in this paper is a system constructed from simple communications primitives. We begin by considering how such a nonlocal pointer comes into existence in the first place. Clearly, it must be communicated by a transaction, on the node containing the object, to a transaction, on another node, which stores it. In its simplest form, this requires a method of transmitting the pointer in a message and re-establishing it as a pointer on the receiving side. Again, this could be provided as a service of primitive communication, but that wouldn't help the problem of using the pointer.

As there must already be a communications abstraction in use, it will be used as the foundation of a structure for handling remote pointers. When a pointer is to be transmitted, it is given a tag (which should be difficult to forge, etc.), the pointer recorded locally with the tag, and the tag transmitted. On the receiving end, a local pointer is created and associated with the tag. In order to make operations on local pointers transparent, the local surrogate object pointed to will invoke a procedure whenever an attempt is made to use a field in it. (This is a primitive capability present in neither PS-algol nor CPS-algol). This procedure may be thought of as an exception handler, and will initiate whatever communication and processing is necessary to satisfy the use of the pointer. A simple example is to obtain a local copy of all or part of the original object using a remote sub-transaction, and then use a post-action to send it back upon commit.

Management of these nonlocal pointers would be simplified by an 'epilog' procedure to be called when the surrogate object was about to be garbage collected. The network node containing the original could then be informed that this reference to it had disappeared. Note that the local association of the tag and the pointer prevents the original object from being garbage collected while there are still nonlocal references outstanding, and that system integrity is not placed at risk by dangling remote pointers anyway, as what could potentially dangle are application created tags.

An abstract object can also, using lower level transactions and communications, maintain an extension of itself on one or more other nodes. Information is exchanged in the same way as for high level transactions. In this model, the object is a higher level abstraction than the network itself, eliminating the need (in this context) for remote transactions, but using them as part of the implementation.

6. SUMMARY AND FURTHER WORK

The principal theme of this paper has been the construction of highly flexible transactions and data objects, using only very primitive system-provided concurrency facilities. The transactions are both atomic and reversible with respect to their effects on shared objects. The model on which this work is based is highly decentralized, distributing the implementation of these properties among both abstract objects and transactions. The principal responsibility, particularly for data integrity, lies with the objects themselves, and is carried out using the post-action facility provided by transactions.

The same basic model has been applied to manage various relationships among transactions. The serialization technique used in the objects was extended to provide dependent sub-transactions. It has been shown that different levels of abstraction and implementation correspond to various degrees of dependency among related transactions.

Also explored briefly has been the application of this model to manage inter-transaction communications of various kinds. It has been suggested how both local and remote communications could be handled, using transaction post-actions to implement the desired coordination when the transactions complete. Methods of implemention have been pointed out for distributed transactions and objects, and objects shared across a distributed network.

The continuation of this work will follow two main directions. One will be to elaborate the implementation of objects and transactions along lines suggested in the paper. Larger and more complex abstract objects will be developed to demonstrate the principles. Particular attention will be paid to communications and distributed systems.

The second focus will be to attempt to formalize properties of objects and transactions and develop rules of induction to propagate properties up through the implementation of abstractions. This has been done here informally, using critical regions to implement transactions. Finally, the invariants of objects and the post-conditions of transactions could be directly related to the formal properties that can be shown for these abstractions.

REFERENCES

[AM84] [AB85] [ABCCM84] [BDMN73] [BG80] [BG83] [BRIN75] [BSW79] [DATE83a] [DAVI78] [DIJK68] [EGLT76] [GR83] [GRAY78] [HOAR74] [KR81] [KRAB85] [LAMP81] [LISK84] [MOSS81] [PS85] [REED78] [WEIH85] [WL85]

Chapter 15

ADDRESSING MECHANISMS AND PERSISTENT PROGRAMMING

W.P.Cockshott
Memex

ABSTRACT *The question of addressing mechanisms is at the basis of all implementations of persistent programming. What persistent programming enables us to do is store complex data structures by virtue of being able to preserve referential structure. What I want to look at is what different forms of addressing can be used to support persistence.*

1. BASIC PARADIGM

The term object oriented programming has acquired specific meaning recently due to the popularisation of the Smalltalk language. When I use the term object based computation I am using it in a slightly weaker sense, more in the way that Intel and IBM use the term.

By object based computation I mean one in which the universe is made up of a conceptually infinite number of objects. Each object has a unique name : its Persistant IDentifier (PID). Given a PID it is possible to do three things:

1. Read a field of the object, an operation with functionality:
   Read(PID, offset -> word)
   Where offset is taken from some subrange of the integers and word is whatever the machine word length may be.

2. Write a field of the object, an operation with functionality:
   Write(PID, offset, word)

3. Start executing code within an object, an operation with functionality:
   Goto(PID, offset)

It is obvious that this is just an "object flavoured" Von-Neumann machine and thus has at least the generality of the Von-Neumann model. It is however much better suited to a whole class of higher level languages than the raw Von-Neumann model. Smalltalk, Lisp and PS-algol, for instance all use this as their underlying implementation model, even if they are usually forced to emulate it in software on a conventional computer.

In the higher level languages the PIDs appear in the guise of some range of data types (pointers, pictures, vectors, strings etc in PS-algol), but beneath the linguistic veil the same set of operations are supported. The question at issue is how do we support the abstractions that characterise the model.

2. CATEGORISATION OF APPROACHES

I will give here a brief run through of the main approaches that are found either in practice or in thought experiments. At this stage I am not concerned with establishing how the different options would perform.

2.1 Textual Flattening

For languages that have a literal representation of every data type it is possible to implement a form of persistence by converting between an in core data structure and its textual representation and back. For languages that support circular structures or are not referentially transparent this is less suitable. This method has been used in Lisp and B [MP84].

2.2 Core Dumping

A simple way of providing persistence is to make PIDs just machine addresses and dump the whole core at the end of a session and reload it at the start of the next session. This method has been used in early Lisp systems.

2.3 Use of paged Virtual Memory

A slightly more sophisticated technique is to make the PIDs virtual addresses in a paged store. It is no longer necessary to dump or reload the entire image. Instead, it can be done incrementally a page at a time as needed. The Symbolics people use this approach.

2.4 Segmented Virtual Memory

There are a number of ways in which a segmented virtual memory may be used for persistence, but the two most significan not distinctions seem to be between systems in which there is one object per segment versus several objects per segment, and whether or not the segments are themselves paged. The IBM system 38 [SOLT81] and the Smalltalk [KAEH81] system on the Alto were variants of this.

2.5 Associative PID addressing + Paged Virtual Memory

This is the technique used in the Poppy experiment. It puts another level of addressing above virtual memory. The PIDs here are names of objects rather than addresses. A combination of associative memory hardware and firmware maps these names onto paged virtual memory. The mapping is dynamic and partial, in that there is no fixed relationship between a name and a Virtual Address and not all names are currently mapped into Virtual Memory. This technique will be examined in more detail later.

2.6 Multiple Address Space Models

The current PS-algol systems [ABCCM84] [ACC83a] [ACC83b] use a multiple address space model (in our case dual). In these the PID is implemented behind the scenes by different addresses according to what address space the object containing it is currently resident in. If it is in core then the PIDs are converted to RAM addresses, if it is on disk then the PIDs are represented as disk addresses. There are two subvariants of this:

2.6.1 Indirect Ram Addressing

In this type of system, a PID when resident is represented as an index into a Local Object Table, the entries in which themselves hold the RAM address of the object.

2.6.2 Indirect Disk Addressing

In a sense any system built upon filestore uses indirect disk addressing, but over and above this, the "disk addresses" used to represent may be entries into a (possibly hashed) table of file addresses.

2.7 Criteria of comparison

In looking at the advantages and disadvantages of the various approaches we will examine them under the following criteria:

a) Simplicity
b) How well they scale up
c) Efficiency in use of physical resources
d) Range of object sizes supported
e) Possibility of supporting transactions
f) Garbage collection and space recovery
g) Support for multithreading and sharing
h) Support for networking
i) Generality

2.7.1 Textual Flattening & Core Dumping

These are the simplest techniques to implement. Of the two core dumping is perhaps the simplest if you have a machine like the Dec System 10 where saving and loading core images is a supported operating system command. As against that the Textual Flattening approach will probably use routines that are already there for i/o or debugging purposes and is therefore portable across operating environments.The problem with both these systems is that they will not be able to hold a collection of data that is larger than can be packed into RAM, since they assume that the whole collection of data is loaded into RAM at the start of each session. This also imposes considerable startup and closedown costs since you must wait for the whole image to swap in or out at the start or finish of the session.

The maximum scale of these systems is therefore dependent upon how much RAM you can afford to dedicate to running your persistent system. This means that the scale of system that they can support grows exponentially with time due to the falling prices of RAM. The bottleneck is then likely to be the time to load and unload the RAM. For long interactive sessions we might be willing to put up with maybe 100 seconds startup and closedown time. The crucial factor then becomes how fast you can transfer data to and from disk at these instants. The theoretical rate of delivery of current disk systems is of the order of 1 megabyte per second per drive. Since it should be possible to store core images in contiguous files transfer rates should approach this. We therefore arrive at the conclusion that core dumping is viable as a technique up to about 50 megabytes, just taking into account disk transfer times.

Given that 256K rams currently sell at about $4 in bulk, giving a raw component cost of $6250 for that much RAM, that is probably somewhat too expensive for a single user workstation, but it is not unrealistic to think in terms of 10 meg systems with present ram technology and the 50 meg systems with the ram technology available within the next 4 years.

In terms of scale therefore these simple techniques remain viable for the sort of application that they were originally applied to: single user interactive programming.

Core dumping probably gives us pretty good use of physical resources within the range of sizes to which they are applicable. It gives us very efficient use of disk store, as data is held in contiguous files. It demands negligible cpu overhead to load and store the data, and no overhead at run time. Programs run at maximum store bandwidth. If one is using only a part of the data in a session then it is wasteful to have all that RAM about, but if the RAM is affordable one might as well use it.

Textual flattening is less efficient in that it demands a considerable CPU overhead on loading and storing, and may result in a less dense representation on disk, otherwise the same considerations apply.

When it comes to support for transactions, they can support transactions at the session level of granularity provided that the file from which the core was loaded is not overwritten when saving the core. They may also support transactions at an arbitrary level of granularity in the high level language superstructure, as shown by the undo facility provided in some lisps.

With Textual Flattening & Core Dumping the problem of garbage collection and space recovery is simple. It exists only at the level of RAM. Only one type of garbage collection has to be carried out. Whether compacting techniques are used or not depends upon whether the ranges of sizes of objects to be used is such that fragmentation is a problem at allocate time.

Implementations based upon these techniques have tended to be single user. It is perfectly possible to implement a multi threaded system that runs in a single core image. It is more difficult to provide a multi user system since loading and storing occur in a single act. It might be satisfactory if the RAM were battery backed up for long enough to allow it to be dumped to disk since one would then only have to dump it occasionally (at 3am for instance). This would only require a few minutes of battery power for the computer.

With conventional machines there would be no hardware protection of data belonging to different users if they all shared the same core image. This would constrain the system to languages where uncontroled assignment does not exist.

A disadvantage of Textual Flattening is that it has no generality, the saved form of data and also its run time representation tends to be highly language specific. Core Dumping on the other hand has some generality in that the technique itself can be widely applied, but the data once stored is unlikely to be shareable because of the difference in the way different languages are represented in core.

2.7.2 Use of Paged Virtual Memory

Given that you have some paging hardware the conceptual simplicity of using paged virtual memory to implement persistence is appealing. In principle is should be possible to build an operating system based upon a one level store that is far simpler than a conventional system that has to manage both a paged virtual memory for programs and a file store for persistent data. The problem is that most machines with paging hardware come with an operating system supplied. This operating system obscures the hardware and may not have enough "hooks" in it for experimenters to be able to use the paging system, so one is tempted to ignore the virtual memory and treat it just like ordinary RAM.

Paging techniques allow virtual memories of almost arbitrary size provided we are not wedded to particular implementations of paging. There are broadly two approaches to implementing a paging system. One of them takes a virtual address and tells you where it is resident or that it is not resident by looking up a page table. The other keeps a table of what virtual page is in each physical page and compares this with virtual addresses when memory requests occur.

The first technique scales up with difficulty, in that the size of the page tables is a linear function of the virtual address space. With very big address spaces this means the tables can not be held as a resident array but must take the form of a trie. As the address space grows the depth of the trie grows logarithmically and along with its depth grows the access times and the software complexity of maintaining it.

Provided that the paging system uses the other technique and just keeps an associative index of which virtual pages are resident, the virtual memory can be scaled up to any size that we can provide backing store for.

With large paged virtual memories the main efficiency constraints to implementing persistence have to do with fragmentation. Since in this system PIDs are just virtual addresses with time the active data gets spread out through the virtual address space leading to less efficient use of physical memory. This makes the system run slower because more page transfers occur and because you may have to chain down a free list of growing size to allocate new items.

To overcome this, compacting garbage collectors of increasing complexity have been devised which aim to keep the working set of objects within a working set of pages that can be resident in RAM.

This reliance upon garbage collection imposes an additional software overhead on performance (the CPU spends part of its store cycles doing compaction) and more seriously ties the system down to the representation in store used by some one language as the garbage collector has to know where to find pointers in objects.

In principle one can have objects of any size you want in a heap in virtual memory. Provided that the technique you use is the same as would have been used in physical memory, it will work in VM. But it will work with degraded performance once the size of your heap gets above your allowed working set of pages. If the mean object size is much less than a page, and if the objects used in a given operation are widely spread over the VM then we will be touching pages that may have only one one object in them. If the number of objects touched in an operation is small enough this does not matter, but ones it exceeds the number of pages in the working set then we will be performing transfer to and from disk that would have been unnecessary with a segmented system.

With paging the amount of physical RAM used can be as much as one page per object unless we have good clustering algorithms. With segments, no more physical RAM is required than is occupied by the object working set. But this advantage of segmentation comes only to those systems that have so many segments that they can afford to assign one per object. Systems with an inadequate number of segments have to be treated similarly to paging systems and have the same disadvantages.

Paged Virtual Memory is particularly suited to implementing transaction systems. Paging hardware will usually keep track of which pages have been modified. This allows the implementation of shadow paging algorithms. The PS-algol implementation on Vax VMS [unfortunately unreported] used this technique.

There is a granularity problem in the use of paging for transaction systems. The page forms the smallest unit of locking and the smallest unit of transaction recovery (by taking before images of the page). Paged virtual memory systems provide relatively poor support for object sharing as the convention is to allow a separate virtual address space to each process. If we then map our objects onto this space we have a set of island universes that can not communicate with one another, the references in the address space of one process are meaningless in the space of another. We could try to get round this by making a certain range of the address space shared between processes (the lower half for instance). This would enable certain designated objects to be shared, but would have the disadvantage that one would have to decide that they were shareable when they were created, and put them in the appropriate area.

Another problem is that paging systems usually protect data belonging to different processes by restricting access to the data in the shared area of VM either to read only, or allowing certain pages to be accessed in write shared mode. This protection paradigm does not map well onto the requirements of an object based system where protection is basically a matter of accessibility, you deny access to an object by encapsulating it within an abstract data type. What you are allowed at is dynamically determined. If you have the PID of an object then you are allowed to get at it. In most paging systems there is nothing to prevent you synthesising addresses and thus bypassing the protection.

Particular languages may prevent this type of cheating, and with them it is possible to have all the processes running in a single heap in virtual memory. But what we gain in protection here we loose in generality by tying the system down to a single language.

The use of paged VM to provide a single level store has proved to be a very practical technique of providing a language independent persistent store. The EMAS operating system, which provides the general user service at Edinburgh University has provided it for over a decade. Any language running on EMAS can use the facility by means of a technique known as mapped files. A designated file can be mapped into a range of the address space where it appears to the programmer as an array. This technique gives an untyped access to the data. It is widely used by hackers on the system, but unfortunately it has not been properly integrated into any language. The conclusion that I draw from this is that a paged VM can provide a language independent persistent store if you are willing to forego type rules.

2.7.3. Segmented Virtual Memory

Segmentation comes in both simple and complex forms. The simple form, used for instance on the Intel iAPX286 micro divides the address space into a set of segments. Each of these may be either present or absent from physical memory. They differ from pages in that although they have a maximum size, they can be smaller. In addition there are often more types of protection built into a segmentation system than in paging systems. The more complex paging systems allow the segments themselves to be divided into pages e.g. ICL system 39.

The important thing to recognise is that segments are fixed partitions of the virtual memory. The implementer of a segmented system then has to decide what to use the segments for. We looked into this when we considered implementing PS-algol on the Intel iAPX286. The alternatives were either to have one object per segment or to fit a lot of objects into each segment. One object per segment is conceptually the nicest scheme, but has the disadvantage of limiting you to a small number of objects (iAPX286 segment numbers are 16 bit). If you put a whole lot of objects into a segment, then you give up some protection advantages that you might have gained from the segmentation.

We thought that the best scheme was to use the segments in a way analogous to the pseudo classes in the Smalltalk Ooze. The objects in a segment would all belong to one class thus allowing hardware to enforce the type rule protection of PS-algol, which in essence only prohibits access to objects whose class definition is not in scope.

In the earlier systems, we have assumed that the allocation of store on disk is a direct image of, or at worst a flattening of the organisation in RAM. On a segmented system this is no longer the case. Since segments vary in size an efficient use of disk surface requires a system that will allocate segments to areas on disk of sufficient surface to hold them. This adds an additional level of complexity to the system. The complexity is analogous to that of a filing system, except that the demands placed on it are also like those on a record based DBMS. It has to handle far more objects than most filing systems have files, yet the individual objects have a greater range of sizes than is usual with DBMS records. The work of writing such a system is a sufficient overhead to deter many implementers from progressing beyond the simpler systems already examined.

The reason that we chose to put lots of objects into each segment in our iAPX286 design study, was because we did not have enough segments. To be able to allocate objects to segments in this way requires a large number of segments. A 32 bit segment number might be enough, but 16 bits is certainly not.

Given that a segment is a fixed partition of the address space, and given that we want to be able to accomodate a large range of object sizes from CONS cells to bitmapped images.

$$A = Log2(N) + Log2(O)$$

Where
N = number of objects supported
O = maximum number of words in an object
A = length of address in bits

For individual object sizes above about 64k it seems sensible to make the segments themselves paged.

One of the traditional advantages of paging over segmentation as a virtual memory technique has been that paging avoids fragmentation in RAM. If we put a heap system with varying sized cells on top of the VM then the fragmentation problem re-emerges at the virtual memory level, thus nullifying one its main advantages. The advantage of segmentation is that we deal with the fragmentation at the level closest to the physical resource. Performing compaction in a paged heap is likely to lead to page faults, doing the same with segments in RAM the operation can proceed without disk transfers, thus leading to a more efficient use of the available memory bandwidth. It also allows us to perform compaction without having to overwrite the pointer fields within objects, since all that is required to relocate an object is to change the segment table. The module responsible for store compaction is therefore language independent.

Again, when it comes to implementing transactions, the advantages of segmented VM depend upon whether there are enough segments to go round. If every object can be a segment then it should be possible to use the memory mapping hardware to keep track of which objects have been modified and by taking appropriate before images to be able to recover the system state at the begining of the transaction. Whereas paging is likely to be rather profligate in its use of the before image file, segmentation allows finer granularity of transfers to this file.

If we assume our garbage collection is to be by pointer following, we need to be able to locate all of the pointers in a segment. This can be done in a number of different ways depending upon how much hardware support we are willing to provide.

One technique is to tag all pointer words with an implicit extension bit that is unreadable to ordinary software, but which can be accessed by the storage management system.

Another technique, used on the Intel 432 [ORGA84] is to divide segments into two types, those made entirely of pointers and those made entirely of non-pointers. The type of a segment would be entered in the segment tables of the machine.

These techniques can provide hardware protection against the illicit forging of pointers and thus of access rights to objects.

The last technique is to rely upon software to use standard conventions as to where it puts its pointers. For instance in PS-algol each object has a header that that describes how many pointers it contains and where they are. This is acceptable if one can be sure of validating all compilers that run on the system. Validating in two senses. They must not allow any illegal highlevel operations on data structures containing pointers. Secondly, the implementation must be certified not to erroneously allow them at a lower level. This is likely to rule out many popular languages from running in a persistent environment, with the consequent loss to you of most of the software that has ever been written. A decision to go this way is acceptable in research environment, but few commercial operations would contemplate it.

Multiple language access is one sort of sharing. The other is between concurrent processes. With a segmented system that uses one segment per object, we could theoretically have a single allocator of segment numbers that is shared by all processes. This ensures that when a chunk of data is passed from one process to another any references to segments in it will preserve their meaning. This makes the sharing of data very straight-forward to implement.

One can envisage a segmented system where the segment number was extended to incorporate the node or processor upon which an object was created. The effect of this would be to provide a uniform object address space over the network. One would still have to decide upon the semantics of shared access to objects on different machines, but generally there would seem to be two approaches to the problem.

One would be to propagate all read and write operations. When you tried to read or write to an object in a remote machine a read or write request package would be sent to the remote machine and the updating would occur there or the result of the read would be returned to you.

Alternatively you could "borrow" the segment from the machine that owned it for the duration of the transaction. This raises questions about locking protocols and the like that can not be dealt with here.

Segmented virtual memory is a very general mechanism provided segments are cheap and plentiful. It makes no assumptions about the programming language used in the layers above because it actually builds the object model into the machine hardware.

The system is only fully general when there is some hardware distinction between pointers and other data objects, allowing garbage collectors to be implemented below the language level.

2.7.4. Multiple Address Space Models

To date all of our PS-algol implementations have been based upon a multiple address space model of persistence. Instead of an object being uniquely identified by a single identifier, an object will have 3 or 4 different types of identifiers that are used to access it in different contexts. To the user of the PS-algol system, it still presents a seamless interface. Objects are identified by variable names, structure field names or array locations, what these things actually represent is hidden from her.

The PS-algol persistence management scheme has to deal with at least three and sometimes 4 different ways of refering to an object. The two lowest level representations are disk and RAM addresses. A disk address is a byte offset into a file, a RAM address is whatever the Unix or VMS virtual machine defines it to be.

Above this we have two abstractions of disk addresses and RAM addresses : PIDs and LONs. A PID is a token for an object on disk and a LON is a token for an object in RAM. There exists a software mapping system for each of the stores that translates between tokens and store locations.

In some of the PS-algol implementations, the map

$$(LON -> address)$$

is degenerate and LONs are simply addresses.

When an object is on disk all pointers are represented by PIDs. When an object is in memory its pointers can be either PIDs or LONs.

I will first explain how the system works with degenerate LONs. When an object is brought into virtual RAM from disk, all of its pointers are PIDs. When an attempt is made to dereference a pointer a software check is made to see if the pointer is a PID. If it is, then a search is made of a memory resident table called the PIDLAM. This holds a two way mapping between PIDs and LONS (in this case LONS are just addresses). Every object that was brought in from disk in this session must have an entry in the PIDLAM.

If it is found that the PID is already entered in the PIDLAM the pointer is overwritten with the corresponding object address and the dereference operation continues.

Otherwise, the PID is looked up in a disk resident index to find the location of that object on disk. The object is then moved into virtual RAM, the PIDLAM updated, the pointer overwritten and the instruction restarted.

When an object is to be transfered back to disk, its pointers are scanned, if any of them are LONs then the LON is looked up in the PIDLAM. If it refers to an object that has been brought in from disk then LON is replaced with the corresponding object's PID. If the object was newly created this session, then a new PID is created for it, an entry is made in the disk resident index and in the PIDLAM and then the pointer is replaced with a PID.

After all LONs have been translated the object is moved onto disk. In the degenerate case where LONs are addresses, it is theoretically necessary to find all pointers in RAM that refer to the object being swapped out and translate them from RAM addresses to PIDs. As it happens we have repeatedly baulked at the complexity of this and have adopted the strategy of completely purging the RAM of all objects at the same time to avoid this complexity.

With non degenerate LONs the system is slightly different. Here the LONs are implemented as indices into the PIDLAM. When an object is brought into RAM all the PIDs are immediately translated into LONs. For those PIDs that refer to objects that are already in virtual RAM the PIDs and corresponding LONs will already be in the PIDLAM. For objects that are not yet in RAM, new LONs are invented and the PIDLAM marked to say that the object is not yet in RAM.

On dereferencing a LON the PIDLAM is checked to see if the object is present and if it is not, it is brought in. Otherwise the PIDLAM translates the LON to a local address which is used to access the object.

In this system the transfer of an individual object back to disk is considerably simpler. All its LONs go back into PIDs, its PIDLAM entry is marked as absent, and the object is swapped out. Since all references to the object go via the PIDLAM entry it is not necessary to go round all the heap finding and updating references to the object.

Overall though, it is clear that neither of these schemes is going to win any points for simplicity. A complex body of software is needed to manage the PIDLAM and the overwriting algorithms. Against this software complexity must be set the hardware simplicity of the scheme. It does not rely upon any special segmentation equipment, and has run on a variety of different machines. The multiple address space model relies upon moving data from one address space into another, from disk to RAM as a session or as we call it a transaction, continues. Problems of scale can hit us in two ways:

1. There is not a big enough address space to handle all of the disk data.
2. Too much data is touched during a single transaction.

Largely due to the influence of IBM the standard word length for a mainframe or super-mini computer has been set at 32 bits. This is enough to address 4 gigaobjects, on most computers these objects are 8 bit bytes but in a multi-address space persistence system, the objects are likely to occupy several bytes. In a Lisp system they might be mainly 8 byte cons cells, on a PS-algol system the average size would probably be a few tens of bytes. In practice multi-address space systems may have to reserve a bit to distinguish between PIDs and LONs, giving us 2 gigaobjects. If we multiply this by the average object size we find that 32 bit pids are capable of addressing between 16 and perhaps 100 gigabytes.

This falls in the same range as the largest amount of disk store on a mainframe. For instance, the ICL series 39 machines are supposed to be able to take up to 80 gigabytes of disk store. It is apparent that if one attempted to replace conventional filing systems with a

PS-algol type persistent store we might be unable to address enough disk store on a large mainframe.

A more immediate problem of scale is associated with the amount of data that is to be handled in a single transaction. As the transaction continues more and more data moves from disk onto the heap in virtual RAM. This starts to cause problems once the heap gets bigger than the physical memory allocated to the process. At this point one starts to get thrashing. If a practicable scheme was available for transfering individual objects back to disk the size of the heap could be held down. As was pointed out earlier, this is very difficult in a system that uses degenerate LONs.

Systems that use genuine LONs, and address objects through the PIDLAM can swap objects out more readily, but the gain in scale achieved by this is likely to be modest. The problem is the size of the PIDLAM itself. When using non-degenerate LONs, we have to keep PIDLAM entries for objects that may never actually be brought into RAM and for all the objects that have been transfered back to disk. The size of the PIDLAM itself is thus likely to pose a problem.

Multiple Address Space Models have a certain amount in common with the core dumping. They share the same basic idea of copying everything back to disk at the end of a session. They differ in that the core dumping scheme must load the entire database into RAM at the start of a session, whereas the multiple address space model does it incrementally.

If we are only going to touch a small portion of the data in a transaction, then moving the data in incrementally seems a good idea, but it must be born in mind that the data that is moved in incrementally will not be stored contiguously on disk. The efficiency of disk transfers is likely to be low with the sort of random seeks that will be requested by the incremental scheme. If a core dumping scheme operates with contiguous files, then one would expect that there would be a range of database sizes and transaction sizes, in which this simpler scheme will be faster.

The multiple address space model places the highest demand on the CPU of any mechanism with the possible exception of textual flattening. In our experience, it takes a considerable optimisation effort to produce a system whose transfers to and from disk are not bounded by the CPU overhead involved.

The multiple address space model can in principle support any size of object up to the maximum that will fit onto the heap. But there is the drawback that with very big objects just as with small ones, the whole thing has to be loaded into the heap before anything can be done with it. This involves a certain lack of efficiency in dealing with large objects that are only partly modified, but is an efficient way of dealing with large objects that are going to be modified or examined in their entirety.

In the multiple address space model, a transaction is commited by copying from the RAM address space to the disk space. In order to make these actions atomic, we use either a shadow paging algorithm for the disk space or a before look file.

If we were to extend this to cover nested transactions, it would be necessary to have N+1 distinct address spaces to allow a transaction depth of N. Given that we have to be able to discriminate between addresses that refer to different spaces, we would have to reserve Log2(N) bits of our address to distinguish between transaction levels.

With these provisos, it should be possible to have a recursive implementation of the multiple address space algorithm that would allow transaction nesting.

Garbage collection in the multi address space system consists in carrying out seperate garbage collections for each of the levels. This is a boon because most garbage is made up of objects that have a short life : the results of string concatenations, function applications and the like. To get rid of these we only have to garbage collect a modest sized heap, 500 kilobytes or so.

It is only the smaller portion of the junk that gets out to the disk that requires a garbage collection of the whole persistent store. This allows these to be made more frequently.

For the RAM resident portion of the garbage collection we have found compaction essential. For systems with degenerate LONs this requires complex garbage collection algorithms that know all about internal data formats. The fact that objects migrate during garbage collection also makes an efficient design of the PIDLAM more difficult. The PIDLAM is a two way mapping table, mapping between RAM addresses and PIDs. The PIDs never change so that a simple hashing method provides efficient lookup, the addresses do change however, so that indexing them is a problem.

Intimate sharing of data between processes is difficult, at least if one wants to go on using the operating system to allocate processes. The point is that only one of the address spaces is shared between processes : the disk address level. This puts us back into the conventional paradigm whereby file store is public but virtual RAM is private.

We can with an interpreter simulate a multiple process PS-algol environment within a single operating system process, in which case all PS processes share both address spaces as they are both running within the same operating system supported virtual memory. An experiment of this sort has been mounted by Larry Krablin, but it was only possible because he in effect abstracted from the existence of multiple address spaces.

If we are willing to put up with the conventional operating system model of shared filestore, then the multiple address space model does allow data sharing at the file level. Our most recent PS-algol implementation on Unix splits the PIDs up into subranges corresponding to different Unix files. Objects with these different ranges of PIDs go into different files.

This solution is conceptually rather unsatisfactory because the files, (thinly disguised as "databases") are an alien intrusion into the PS-algol language, and there is no clear rationale as to why an object lives in one file rather than another.

It is conceivable that one could operate a networked system based upon the multiple address space model but the conceptual complexity involved would be considerable.

Sharing of data on one machine is possible in our current implementation because all users share a common disk address space of PIDs which are mapped onto Unix files. A relatively simple extension of our system would be to further subdivide the PIDs, so that subranges of the PIDs would correspond to files on different Unix machines linked under the Newcastle Connection. This would work for small networks, but if they were scaled up then we would run short of bits in our 32 bit PIDs to distinguish machines.

To be able to access networks or arbitrary size we would need some scheme whereby objects on remote machines were refered to via alias objects on the local machine. The alias objects could contain the whole network path to the remote object.

It is said that the worst mistake in computer architecture is not to provide enough address bits. Whenever this mistake is made by the original designers, subsequent users have to go through contortions to get round it. Attempting to fit persistent addressing into the 32 bit limit originally designed to be ample for RAM addressing is such a contortion.

Multiple address spaces have comparatively low levels of generality. It is possible for more than one language to use a multiple address space system provided they both stick to a common definition of the format of data objects on the heap. However, the intimacy of the connection between the persistence management system and the language's conventions on data formating seem to violate the principles of modular design, and make it doubtful that multiple address space models will be a viable method of sharing data between different programming language environments.

	Persistent Object Space	
Objects	Operations	
PIDs	Read(Pid,Offset -> word)	
Offsets	Write(Pid,Offset -> word)	
	Words	
	Persistent Virtual Memory	
Virtual Addresses	Read(Virtual addr -> word)	
Words	Write(Virtual addr -> word)	
	Physical Ram	
Physical Addresses	Read(Physical Addr -> word)	
Words	Write(Physical Addr -> word)	

Figure 1

3. WHERE NEXT ?

If persistent programming is to get beyond the stage of being an interesting feature of particular programming languages we will have to develop implementation approaches that are language independent and which support in a consistent way both those requirements that have traditionally been met by filestore and the new needs of heap intensive programming.

It is OK for experimental systems like Smalltalk or PS-algol to create isolated self-sufficient environments, but to become of general relevance to the computing community we must provide persistent data that can be shared between users, languages, and machines as readily as files can be at present. The persistent programming environment will have to be able to give ready access to the existing collections of data and programs on systems like Unix. This will involve paying much greater attention to what we may have considered to be operating systems or machine architecture problems. We need a model of persistence which can represent and live alongside Unix.

3.1 Implications for Machine Architecture

A lot of the difficulties we have in implementing persistence come from our having to use machines and operating systems that were not designed with it in mind.

The most suitable machines for supporting persistent objects are those with segmented architectures. As I have argued above, there are two limitations to most segmented machines : the number of objects that they can support, and the range of segment sizes that they support.

If one is to support persistent addressing over large networks of machines, it is necessary to be able to be able to construct an unambiguous reference to any object on any machine. This is potentially a very large number of objects. Luckily the number of objects you can address is exponential in the number of bits you use, so a quite modest extension of address spaces would give this capability.

Back of an envelope calculations indicate that if one is thinking just of one computer a 48 bit object (segment) number is likely to be enough. A computer creating half a million objects per second would take over 10 years to run out of object numbers, by which point it is likely to have become obsolete.

To identify which computer a data object lives on it seems likely that about the same number of bits would be needed. This figure can be arrived at in two ways : the number of bits needed to encode an international phone number, or the number of bits in an ethernet address. Both of these standards for world wide addressing contain substantial redundancy, but that is probably inevitable in the absence of a centralised allocator of addresses.

This indicates that segment descriptors should contain at least 96 bits. Since we computer numerologists have a deep faith in the arcane significance of powers of two, it seems a good idea to round the number of bits up to 128 and use the spare ones to describe the object size and type. These 16 byte descriptors are the same size as those used on the IBM system 38 and only twice as long as the ones with much more limited capability provided on the ICL system 39.

There are probably a number of different engineering solutions that are possible to get over the problem of handling both lots of small segments and smaller numbers of larger segments. I will outline one possible solution just to show that the problem is tractable.

3.2 Associative Object Addressing

I assume familiarity with ordinary paged VM. I propose an additional layer of addressing, object addressing, that will sit on top of paged VM, as shown in fig 1. Objects are addressed using a segment descriptor and an offset. Associative hardware maps the segment descriptor to a shorter paged virtual memory address. An investigation of the problems involved with translating large segment numbers has shown that the translate time is dominated by the time required to read the segment descriptor from memory [COCK84], and thus that the cost of dereferencing a pointer is of the same order as assigning to a pointer variable. Consider the problem of translating a 16 byte PID on a machine with a 32 bit bus. Fetching the PID from memory will take 4 cycles. If the PID is then transfered to an associative translator, then 4 more cycles are required to send the PID to the associative memory, and one to read the result back. In addition we have the lookup time for the associative memory, which is likely to be comparable to a bus cycle. On this basis we arrive at a figure of 10 cycles to translate an address, with perhaps one more cycle to fetch the object dereferenced (assuming it to be 32 bit). This is to be compared with 8 cycles to do a pointer assignment, assuming that we are copying between two stack variables: 4 cycles to read and 4 to write.

The hardware techniques used in implementing the associator can be the same as those employed in normal set associative address translation caches. These involve the computation of a hash on the key (address or PID) which is then used to look up a tag in one or more fast RAMS. These TAGS are then compared in parallel with the key or part of the key. If a match is found, another part of the RAM word supplies the datum associated with the key.

The resulting virtual address has the offset added to it to yield the address of the appropriate word in the object within paged VM. The paged VM acts as a conventional single level store, providing an abstraction over different forms of physical store.

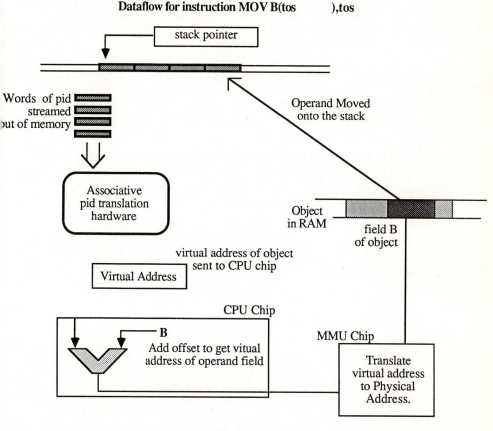

Dataflow for instruction MOV B(tos),tos

Figure 2

This process is shown in figure 2. It will be noted that the PID translate cache performs its translation on PIDs as they are read in from memory, whereas the MMU translates outgoing virtual addresses into physical addresses. This arrangement requires fewer bus cycles than placing the translator on the outgoing address path. In fact it reduces the whole process of dereferencing to 6 cycles: 4 to fetch the PID into the translate cache, 1 to read the result from the cache and 1 to fetch the dereferenced result. This is an overhead of only 1 cycle over the time needed to read the PID and the dereferenced integer.

Three types of virtual address faults are possible with this mechanism:

1 Object faults, when the object either does not exist or is on another machine. In the latter case a copy must be brought into the machine from its domicile by sending a request over the network to the machine specified in the segment descriptor.

2 Page faults, the page is currently on disk not in RAM. The page is brought in.

3 The associative hardware would only be able to remember the virtual addresses of a subset of the total collection of objects. When it cannot remember an address it generates a cache fault interrupt and the CPU has to consult an index held in virtual memory. This index could be based upon some well established technique like linear hashing. It would function like an ordinary PIDLAM and map PIDs to virtual addresses. If this software procedure failed, then a fault of type 1 would be generated.

3.2 Object Placement

With the system described the segment descriptors are best thought of as names rather than as addresses. The objects may move about in virtual memory but always retain the same name.

This allows the dynamic clustering of objects at run time to optimise paging performance. Figure 3 shows a possible arrangement of virtual memory. It is divided into 3 areas:

1 The object directory. This is an index, presumably accessed by some hashing technique that defines where an object is in virtual memory. The most recently used entries in this directory are copied into the associative memory.

2 The Megaheap. This area of paged VM holds the inactive objects plus all the objects that are more than 1 page in size.

3 The Microheap. This area holds all of the active small objects. This Microheap should be smaller than the amount of physical RAM so that the paging algorithm will result in most of it being RAM resident. When a small object's PID is entered into the associative memory, then the object itself is copied from the mega to the micro heap. When an object is pasivated by being deleted from the associative memory it is copied back to the megaheap. New objects are created on the micro heap. The objective of the Microheap, is to concentrate the working set of objects used by a process in a small area of virtual memory. This should reduce the ammount of virtual memory used by the process, and prevent thrashing.

3.4 Garbage Collection

A big advantage of using long associative PIDs to get at objects, rather than addresses, is that they contain redundant information. The number of possible PIDs is astronomical, but the number of objects resident in the micro or even the mega heap is much smaller.

As a result the garbage collector need not know about the formats of data objects. It does not need to know where the pointers are. It can just scan through objects and assume that each 4 word sequence is a PID. If it is one, and the object is in the micro heap, then the PID will be found in the associative memory. If it is not a PID or not the PID of an active object it will not be in the associative memory. Thus all pointers will be found, and very occasionally, a random sequence of other data may result in an extra object being retained on the heap that could have been garbage collected.

There are other complexities of the garbage collection algorithm to do with having two heaps, but these can be dealt with using the same techniques as existing PS-algol implementations.

It should be noted that as with the dual address space model, most junk (assuming it has a short life) is discarded during garbage collection in a RAM resident heap. This means that garbage collection should not entail much paging overhead.

3.5 Generality

By putting associative PID addressing into the machine hardware, it becomes possible for the provision of persistent store to become orthogonal to the programming language. PS-algol has demonstrated that the provision of persistence is possible with only minimal changes to a programming language (in our case the provision of a few library routines).

It is easy to see how Pascal could be similarly extended. On a machine of the type described persistent objects from both PS-algol and Pascal could be shared and exchanged.

Languages like C or BCPL which allow explicit address manipulation would need new operators to enable them to dereference PIDs rather than addresses. It may be objected that it is dangerous to allow languages without strong typing to get at persistent data, but I think that this is mainly a matter of taste. There is a thriving culture of C programming in a Unix environment where the only persistent data provided is the completely untyped Unix file.

In this view, the advantage of persistence is not in any type security that it offers, but rather in the fact that it allows complex graph structures to be stored with the same ease that we can currently handle serial sequences of characters. There is a danger that most current work in persistence, since it is posed at the level of programming languages rather than at the machine architecture/operating system level, will result in the birth of Island Universes. Collections of data that are only able to be manipulated by one language and which cannot be readily exported to a different language environment. If the mechanisms for persistence could be built into a popular family of microprocessors, then the Mountaintop Stronghold syndrome could be avoided. A large body of languages and machines by different manufacturers would be able to store and exchange persistent data.

252

3.6 Feasability

With funding provided by Acorn computers I have constructed a prototype machine, the "POPPY" with the addressing structure described. This is a half scale prototype, with 16 bit data paths and 64 bit PIDs. The associative addressing mechanism has been tested and shown to work. A PS-algol compiler for the POPPY is currently underway. Detailed description of the hardware is beyond the scope of this paper, but see "Building a Microcomputer with Associative Virtual Memory", P Cockshott, Glasgow University 1986.

REFERENCES

[ABCCM84] [ACC83a] [ACC83b] [COCK84] [KAEH81] [MP84] [ORGA84] [SOLT81]

Chapter 16

The Implementation of Galileo's Persistent Values

A. Albano[1]
G. Ghelli
R. Orsini
Università di Pisa

ABSTRACT *Galileo is a conceptual language for database applications in which the persistence of values is an orthogonal property, i.e., values of any type are persistent as long as they are accessible from the top level environment. In Galileo providing such property poses difficult problems since the language is based on a heap memory management, with variable size elements and an incremental garbage collection, and it allows user control of failures and undo of updates. The interaction of these features is described and the approach adopted in the implementation now underway is discussed.*

1. INTRODUCTION

1.1 Motivation

Most commercial systems for the management of persistent data must be programmed using a language in which persistent and temporary values are treated in a truly different way. The distinctions concern the following aspects:

- Operations The only operations permitted on persistent values, which are usually records, are searching, which locates one record or a subset of records, reading, which copies the persistent record into a temporary structure, and writing, which copies a temporary structure onto an old or a new persistent record; in other words persistent records are operated on only with a "copy semantics". On the other hand on the temporary structures the complete set of operations offered by a programming language is defined.

- *Program structure* Sometimes the operations on persitent data must be marked to be recognized and managed by a precompiler and inside these markers a different syntax or even different control structures are used.

- *Data structures* Persistent values can be of a limited number of types. For instance, they can only be flat records grouped in special collections, DBMS dependent, among which associations are defined (relations, segments or Codasyl records). For temporary values, instead, a large variety of structuring mechanisms are supplied, such as abstract data types, arrays, sequence of values, tree structures, sharing of values, etc..

---

1 This work was supported in part by Ministero della Pubblica Instruzione

- *Failures management* DBMSs support a transaction mechanism with a recovery algorithm to ensure that the database includes only updates produced by transactions executed to completion. However, the recovery algorithm acts on persistent values and not on temporary values. This choice is a consequence of the fact that DBMSs consider as part of a state only information stored into the database and not the temporary values created during transaction execution. This solution is unsatisfactory when transaction can be nested and a control structure is provided by the language to execute alternative actions when a sub-transaction fails.

- *Memory management* DBMSs offer operators which explicitly remove values from databases, while temporary data could be managed with a stack discipline or with a heap discipline with garbage collection.

As consequence of these different capabilities, programs using both persistent and temporary values are rather complex, because the programmer must write lots of lines whose only task is to move and to convert data structures between temporary and persistent format. To transfer flat persistent records to and from flat temporary records can be long and wasteful but simple, but for graph-like data structure these conversions are expensive and error prone. This drawback is becoming more serious nowadays that there is a growing demand of languages and systems for using persistent and complex data structures, such as those present in AI and CAD applications. As it has been pointed out in [AOO82] [ATKI78] a solution should be found by designing a persistent language, i.e., a programming language with persistence as an orthogonal property of the type system: instead of having persistence as a property of values of special types, it is assumed that persistence is a property of any value that is accessible from the top level environment. Most languages offer different persistent and temporary data types to cope with the different nature of the hardware used to store them, while in a persistent language values are managed at a level of abstraction where these differences disappear. In other words a persistent language offers a very high level of abstraction over the machine and consequently more difficult implementation problems are expected. Other attempts in producing a new language with orthogonal persistence are presented in [ACC81] [ABCCM81] [BROO82] [ABCCM84] [CORM83]. A comprehensive description and analysis of persistent programming languages is contained in [ABCCM81].

1.2 Structure of the Paper

The purpose of this paper is to illustrate the approach adopted to implement the persistent language Galileo [ACO85] [AOO86]. Since the implementation is in progress we can not comment on the goodness of the approach yet, however we believe that it is interesting to discuss the difficulties we are coping with since Galileo has features that make the problem particularly difficult to solve. A description of the language is beyond the scope of this paper and so only the features that affect the implementation of persistence will be briefly outlined.

1. In Galileo it is impossible to know at compiling time whether a value is persistent, and the transfer of values between temporary and permanent memory is not under the control of the program, but the run-time support system must decide where to keep values and when they must migrate from one memory to the other.

2. The memory management is not static, but is based on a heap with elements of variable length. Attention must be paid to the algorithm for garbage collection since there can be a large quantity of values and most of them are in the permanent memory.

3. Galileo offers a mechanism for the nested management of failures and transactions. Expressions can fail and this event can be controlled to perform an alternative action. When an expression fails all updates on any value are undone; consequently, an undo log must be kept for any modifiable value. The garbage collector, in order to establish that a value is no longer accessible, and so it can be deallocated, must take into account the possibility that the value may become accessible later on in the case of an expression failure.

4. Galileo supports the abstraction mechanisms of semantic data models to model databases: classification, aggregation and generalization. It is also assumed that an element of a class cannot be removed from a class as long as it is used as a component of other class-elements (*dependency constraint*). This assumption has been included to avoid the problem of "dangling references", which arises in relational systems when a foreign key no longer refers to a tuple of a relation because this has been deleted, or its key has been changed. The enforcement of this constraint is not easy because in Galileo it is possible to define circular data structures with class elements.

5. Besides the persistence of user values, another problem is to deal with the persistence of data created by the language compiler, data which may have a complex structure.

The next section presents an overview of the approach adopted to implement persistence. Section 3 discusses the problem of persistence values. Section 4 discusses the recovery algorithm for user failures, soft-crashes and hard-crashes. Section 5 discusses the problem of garbage collection when values with different volatility are present. Section 6 describes an algorithm to enforce the dependency constraint. Section 7 discusses the persistence of compiler data structures and its consequences on the implementation of Galileo. In the conclusions we comment upon the implementation which is now underway and our future plans.

2. AN OVERVIEW OF THE APPROACH

Let us present an overview of the approach to the implementation of Galileo, before discussing every aspect separately in the following sections.

Presently a Galileo expression is compiled to an intermediate code of a Functional Abstract Machine (FAM), a stack machine designed to support functional languages on large address space computers [CARD83a]. The instructions of FAM are assembled into VAX machine code and then executed. The compiler typechecks expressions and then translates them. The data needed by the compiler are essentially the *type environment*, which is a map from identifiers to definitions of types, and the *run-time environment,* which is a map from identifiers to the position of denotable values in the FAM stack. The FAM manages the memory, which is a map from locations to storable values, and the values in the run-time environment, stored into its stack.

To deal with persistent values, the FAM memory is implemented as a "persistent virtual memory", that is it is divided in pages and disk-based; only the pages recently used are kept in working memory. For efficiency reasons addressing space is divided into two parts with different expected behaviour, to focus efficiency efforts on the part where more used data are located. Following the classification in [COCK85] our approach is based on a paged virtual memory with a multiple address space. The FAM memory is managed with the Lorie's shadow pages algorithm to deal with soft-crashes, i.e., crashes in which the content of temporary memory is lost, but the contents of permanent memory (disk) remains intact [LORI77]. To deal with hard-crashes a technique based on incremental dumping is adopted. To recover from expression failures, a log for undoing assignments is used [GRAY78].

For garbage collection an algorithm based on the proposal of Baker and Hewitt and Lieberman is used [BAKE78] [HL83]: the memory is divided in spaces and values are stored in different spaces according to their expected lifetimes. The garbage collector is "partial", i.e., it concentrates the efforts on values most likely to be garbage, and it uses incremental copying garbage collection, i.e., the garbage collector interleaves its operations with those of the user program and copies one object at a time. Objects are copied into contiguous locations so that the virtual memory performance is improved through compaction.

3. PERSISTENT VALUES

To cope with large quantities of data, persistent values must be stored in the persistent memory even during program execution, and only the actually ones used can be kept in the temporary memory.[2] To find out which values must be loaded into the temporary memory two approaches exist:

- Static (or explicit) approach: the language provides some primitive operations to specify that a value must be transferred (or these primitives can be inserted by the compiler, if the lifetime of objects can be statically checked).

- Dynamic approach: the system transfers values into temporary memory when needed. In case of space shortage, some values are selected to be moved back into the persistent memory.

The first approach is the standard one, because of its simplicity, but it is suitable only for languages with copy semantics or, at least, in which temporary and persistent data can be statically recognized. Since Galileo radically departs from this philosophy, this approach cannot be adopted. The second approach allows an homogeneous treatement of persistent and temporary values, and so it is more appropriate for the Galileo implementation.

In any system which manages persistent data, each persistent object is uniquely identified by its address in the persistent memory (PID), which allows a "direct" access to its disk version. When the value is loaded in temporary memory, it gets a temporary memory address, or TID, and the system has to maintain some structures to execute the translation PID-TID when necessary.

When persistent data are needed in the working store they can be separately loaded, each receiving a temporary address of its own, or the pages in which they reside can be entirely loaded, so as to maintain a table of correspondence only for page addresses. The first approach simplifies the interfacing with an existing DBMS, as we can let the DBMS maintain the persistent values, using its own identifiers as PID. The second approach considerably reduces the space required by the tables and could require less disk accesses to compute over many little items of data physically contiguous, so it is more suitable for

---

2 By temporary memory, we mean either real store or virtual store, depending on the actual computer

a system in which each value is persistent, and for this reason it has been preferred in the Galileo implementation.

In some systems, when a PID is translated, it is substituted by the corresponding TID in the cell where it has been found, to avoid translating it again on a successive access. This requires that the correspondence PID-TID should be fixed during a session that it should be possible to locate each translated PID to translate it back to its original value when the correspondence changes. Neither of these conditions holds for the system we are describing, so this technique will not be adopted.

A system which manages persistent and temporary data in a uniform way, manages data structures containing two different kinds of address, PID's and TID's. Every time an address is used if it is a TID the operation is immediately executed, while if it is a PID it is translated into the corresponding TID, and the object is loaded if not already present in temporary memory. Since in Galileo there is no way to distinguish persistent from temporary data, it is simpler to use, at the FAM level too, only one kind of address, PID. This means that no checking must be made to distinguish PID's from TID's, but each address found in a data structure must be translated to be used. If the FAM were implemented at machine level on a virtual memory processor, like VAX or 68020, we would not incur any overhead, as these processors do execute, with hardware support, a page translation for each memory access and provide a mechanism to check the presence of a page in the working memory. On the other side an implementation at this level requires the rewriting of the kernel of the operating system, which is beyond our aims.

The implementiation of the FAM over the operating system poses the problem of avoiding as many address translations as possible. One key observation to avoid translations is that a great part of the computing is performed on the most recently created data. So these data are maintained in a distinguished part of the addressing space, called the Most Recent Generation (MRG), where PID's and TID's numerically coincide, and where objects are guaranteed to be always present in working memory;consequently when an address is in the MRG it does not need to be translated. A few observations can be made:

- This technique makes sense only working upon a virtual memory system, because a large portion of temporary memory is reserved to the MRG; if temporary memory means virtual memory there are no problems, while a physical memory should be managed in a more effective way. On the other side, as already observed, if the implementation is made at the lower level there is no need of this technique.

- The MRG reintroduces, in some way, the PID/TID dichotomy, but a lower level, where it can be used to increase efficiency without affecting the programmer.

- The division of data in generations is made by the garbage collector too. Also the garbage collector must manage in a special way the MRG. This is treated in Section 5.

This schema could be sufficient for the implementation of a language in which functions are not first order objects. But in Galileo functions are objects like any others: they can appear as component of other objects, can be the result of other functions and they can be dynamically built and used. Functions are the critical type of data for the Galileo system, because every time a piece of data is accessed, the corresponding piece of code, usually longer, must be accessed too. But for functions it is not true that most elaboration efforts concentrate on the most recently created, so our MRG schema does not optimize code accesses. This problem is peculiar for higher order functional languages and it requires more investigation.

4. RECOVERY FROM FAILURES

Failures can be classified into three different categories, depending on their effects and on the complexity of the techniques used for recovery:

- user failures: caused by predictable run-time errors (or keyboard generated interrupts) occurred during a program execution or generated by the user with an explicit command; they leave the machine in a well-defined state.

- soft-crashes: unpredictable system failures which do not affect the persistent memory. They leave temporary memory in an undefined state.

- hard-crashes: disk failures.

4.1 User failures

Galileo allows the dynamic definition of nested and selective failure handlers, which trap a failure depending on its nesting and kind. For this reason, at any time the system should be able to restore all the states associated to the currently active failure handlers. Besides, in contrast with usual DBMS languages, Galileo requires the global state to be restored, and not just a subset of it, i.e. the database.

For this kind of failure management, techniques which only maintain a copy of the old state to be restored together with the current version of the values cannot be used, because of the several states that can be restored. An "undo log" has been chosen instead, that is a sequential file of pairs (location, old value), one for each assignment operation executed [GRAY78]. When a failure occurs, all the assignments since the appropriate failure handler definition are undone. If this log had to be kept for assignments to physical cells, its cost would be intolerable. However, since Galileo expressions are compiled into FAM operations, the Log can be maintained only for the values as seen at this abstraction level. Besides, only a small subset of FAM operations perform updates on the machine's abstract memory, and for these operations the log keeps only the address of the updated structure and a pointer to the old value. In addition to this, for most inner loops variables, on which most updates are performed, it is possible not to take any log record, if they are declared in an expression which do not contain code installing new failure handlers.

4.2 Soft-crashes

For soft-crash recovery the "shadow pages" technique has been adopted [LORI77]. For each data page, on the disk a current version and a "shadow" version exist, which usually coincide. As the set of the shadow pages represents a stable and consistent version of the persistent values, a new current page, distinct from the shadow version, is created the first time a page is updated. A corresponding two-versions table exists, which maps logical page addresses on physical slots on disk: the shadow version refers to the shadow pages; the current version refers to the version on disk of the current pages. Since all updatings are performed on the current version, the shadow version of both data and tables is safe from soft-crashes. Periodically, executing the "save operation", the current version becomes the shadow one. This process has two main phases: first, the pages of current version still in temporary memory are rewritten on disk; second, the current version of the page table becomes the shadow version, with an atomic operation, and the old shadow version is discarded.

Although this technique is expensive, when it is combined with the "virtual memory" approach previously described, the additional cost is low. In fact, the major component of the cost of the shadow pages technique is the address translation and the page management, and these tasks are already performed by the virtual memory component. Working on a virtual memory processor at bare machine level the shadow pages technique would not pose any substantial overhead at run-time, and the main overhead would be the rewriting of modified pages and the table switching at save time.

The shadow pages technique is combined with that of the undo log previously described, to form a two-level consistency management: at the abstract level, the log maintains the consistency of the user's transactions, coping with the FAM data structures, while, at the low-level, the shadow pages maintain the consistency of the FAM operations, allowing them to be regarded as atomic operations by the log management. The operations of the incremental Garbage Collector are synchronized with FAM operations, so that shadow pages guarantee consistency of data against Collector's actions too. This is really important because Garbage Collector operates continuously and it not easy to keep trace of its actions.

An alternative technique is that of "Write Ahead Log", which maintains only one version of the machine's state, and keeps a log file complete enough to restore a consistent state, starting from a non-consistent one left by a soft-crash; is a one-level consistency technique. This technique is not very suitable for a language with a dynamic memory management like Galileo. In fact, as it will be shown in the next section, an incremental copying garbage collector is used, and a low-level Write Ahead Log would keep track of all the collector's operations, becoming rapidly intractable.

Another choice to be made is between an undo or an undo/redo log. The first kind of log allows the restoration of the states preceding the last save operation, while the second one also allows the restoration of the states following it, provided that the log is saved more frequently than the state, and provided that the log contains not only the address and the old value for updated cells, but also the new value. We have choosen the undo log for two reasons: first, because the redo log has a greater overhead in terms of space and time, as it even keeps a record for every allocation operation; second, because it interferes with the normal operations of the garbage collector and the allocator. In fact during the redo phase following a crash the redo log tries to determine the reallocation address of objects, conflicting with the usual allocation policy.

4.3 Hard-crashes

Disk crashes can be treated in a way similar to that of the soft-crashes, incrementally saving on a tape a FAM consistent state, and using the log to restore a consistent state from tape information. Periodically, a background process is started to copy incrementally all the shadow pages modified since the last dump. This activity does not interfere with the normal system's operations, since it reads only the shadow pages, and so it can be run concurrently with it. This process can last much time without problems because the shadow pages that it reads can be locked to prevent their reuse. Because of the lack of the redo log, the dump process should be done rather frequently, to have a recent state to restore in case of crash.

5. GARBAGE COLLECTION

The execution of a Galileo program in general requires dynamic allocation and deallocation of large quantities of memory, usually managed through a garbage collector, which finds the inaccessible memory space and makes it available again. The standard algorithm, which visits the whole memory, marking the accessible cells, does not work well in this case because of the very large amount of relatively stable memory involved. A "partial" algorithm is necessary, like the Hewitt's algorithm [HL83], which visits only a small slice of memory, containing the most volatile values. This algorithm is based on the assumption that the values can be divided in a small quantity of highly volatile, short-lived, intensively accessed ones, a very large bunch of stable, long-lived ones, and a variable quantity of values with intermediate characteristics. The aim of this algorithm is to visit only dynamic values, in order to achieve a good ratio of freed cells to visited ones.

In Galileo, newly created values refer only to already existing values; so that stable values, which are allocated before the dynamic ones, do not usually contain references to these. This property of the language can be exploited by the collector, which, in principle, can visit only the dynamic portion of the memory, which should be small. The few pointers from stable to dynamic values, arising from assignments, can be constrained to point indirectly through an entry table, so that the collector, visiting this table, can locate all dynamic objects pointed from the static area. To keep trace of the age of objects, they are divided, depending on the creation time, in "generations". An entry table is associated with each generation, through which pointers from preceding generations pass. Values in most recent generations are frequently collected (Fig. 1).

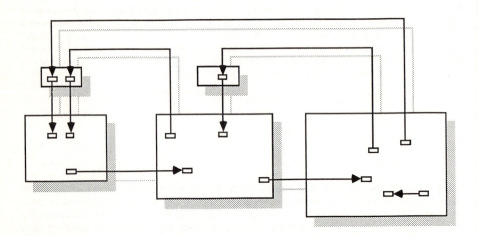

Most recent Generation: here values are created; pointers from this generation do not pass through tables.

Intermediate Generations: the data in these generations have survived some garbage collections.

Oldest Generations: the data in these generations have survived many garbage collections so can safely be considered stable.

Figure 1 : Data Generations (*)

The collection inside the generations can be performed using Baker's algorithm, which copies all reachable values in a newly allocated set of pages (new space), discarding the old set (old space) at the end of the operation [BAKE78]. This operation can be intermixed with those of the user, so that he will not be periodically interrupted by a long garbage collection; this feature requires the possibility of distinguishing between addresses referring to the new space and those referring to the old one, which have to be treated in a particular way. Actually, when an address referring to the old space is found, the object referred is immediately copied in the new space and its new address is used instead of the old; in this way old addresses are guaranteed not to appear in the new space.

This is a linearizing algorithm, in that it moves reachable values to contiguous locations, to avoid fragmentation, which is a severe problem for variable size values, and to increase the efficiency of the virtual memory mechanism.

To cope with data sharing and circularity, the collector replaces values in the old space with their new address, while moving them in the new space, so that the other references to the same object will be replaced by references to the same new version. A crash during this rewriting operation can cause the loss of the value contained in the cell, if it is overwritten before it is moved, or the loss of the link to the new value, if the object is first moved and then overwritten. This problem can be solved either taking log records for collector's operations or making them atomic, with the shadow pages technique. This is the main reason for our choice of the latter mechanism for soft-crash management.

The existence of both a moving garbage collector and an undo log can cause some problems. The garbage collector can interfere with the log's activity in two ways: it could free cells currently unreachable that could be made reachable again recovering from some kind of failure; it could change the address of objects pointed by the log record, making it obsolete. To avoid these problems, the values pointed by the undo log are considered reachable, so that restorable cells will not be freed, and the addresses contained in the log will be updated by the collector. A log record must be removed at the end of the transaction, so that the collector will eventually free its values.

6. DEPENDENCY CONSTRAINT

Galileo allows the simultaneous removal of a group of objects, i.e. class elements. This operation fails if the objects are components of other objects, that is they are pointed to by them, to avoid creating dangling references. A way to enforce this constraint is through reference counters, associating to each cell a counter of the number of cells reachable from objects which directly refer to it; however, it is well known that the standard technique for reference counters management can not cope with circular structures, which are allowed and useful in Galileo [KNUT73]. A new algorithm has been devised to cope with such structures, exploiting the information acquired by the compiler to guess where cycles of references can be [GHEL85]. In the first phase, a subset of values is chosen as the set of "suspect" values. When an object's counter becomes zero, its children have their counter decremented, as in the standard algorithm. In addition to this, when the counter of a suspect value is decremented, its children have

---

* Although only three generations are drawn, a greater number will be used in the implementation

their counter tentatively decremented too. At the end of the procedure, if the counter of the suspect value is not zero, its children's counters are re-incremented (for a complete description see [GHEL85]). If the set of suspect values is chosen according to certain conditions, this algorithm is guaranteed to maintain correct counters even in presence of circular structures. It can be demonstrated that in Galileo it is possible to choose a really small subset of suspect values, easily determined at compile time, to make this algorithm rather efficient. This method requires sharing of information between compiler and run-time support. The compiler can supply this information at compile time, but an exchange of information at run-time would be more convenient for efficiency reasons. The necessity of efficiently associating to each value its type and of access from FAM compiler information poses us some problems which have not been completely solved. We considered an association object-type made at page level, which would be efficient in terms of space and time but would suffer of fragmentation problems, as in each page only values of the corresponding type could be stored. This fragmentation would make more difficult for the garbage collector to predict the space needed for copying a generation, because fragmentation can increase the quantity of memory required to keep the same data reordered by the collector; the possible growing of a generation when copied makes more difficult a static allocation of generations which is essential to an easy and efficient generations management.

7. THE PERSISTENCE OF COMPILER DATA STRUCTURES

As we said in Section 2, maintaining the information about types and identifiers is a compiler's task. This information must be persistent; it is characterized by a complex structure and used by procedures executing complex tasks (type inference and sub-type unification; see [AGOP85]). So the compiler has to be written with a high-level "persistent language" supporting complex data structures.

The compiler is currently written in Pascal. To go on using this language, procedures should be defined to read, write and allocate persistent values for every data type defined, and should be used instead of the Pascal operators "^", ":=" and "new" for allocating and using data structures. The developer has to decide which values are persistent, and to reclaim space which becomes inaccessible, a task that we find really hard. These problems would make the writing the compiler too hard.

A dramatically simpler solution would be given using a persistent language. The best choice would be to use Galileo itself for two kinds of reasons:

- Galileo supports all the data types and operators necessary to write a compiler and it completely takes upon itself the management of both persistence and memory allocation/deallocation.

- The complete homogenity of data structures used by the compiler and the run-time support permits an easy exchange of information, that is needed for the following purposes:

 a) In Galileo when a type is defined, a function is automatically defined allowing the creation of values of that type through "forms" [AFL85]. This is implemented through a general-purpose "generic" routine which uses, at run-time, the description of the type managed by the compiler.

 b) To implement the Remove instruction, as indicated in the preceding section.

c) To query a metadatabase, containing the information on the definitions, which is maintained by the compiler.

The Galileo code implementing the compiler could be compiled to FAM by the current compiler or even by itself. In both cases, as soon as we have a persistent FAM, the produced code would manage persistent data structures. So we do not need the construction of a hierarchy of incomplete compilers as is usually done when bootstrapping from scratch.

8. CONCLUSIONS

An overview of the problems to be solved while implementing the persistence in the Galileo language has been presented. The adopted approach is based on: **a**) a persistent virtual memory technique to guarantee the persistence of every reachable value; **b**) an undo log to recover from users' failures; **c**) a shadow page mechanism, which can be derived from persistent virtual memory with a moderate overhead, to cope with soft-crashes; **d**) an incremental dump performed without interfering with user operations to recover from hard-crashes; **e**) a moving incremental partial garbage collector for the management of a memory characterized by the very large size, with both relatively stable and highly volatile data; **f**) an algorithm based on reference counters, which can cope with circular structures profiting from compiler information, to enforce the dependency constraint; **g**) the use of Galileo itself to write the compiler, to obtain the persistence of compiler's data.

The implementation is now underway, and the goal is to have soon a first prototype to experiment the proposed architecture for single user environment. Later on an extension to provide concurrent access will be considered.

REFERENCES

[ABCCM81] [ABCCM84] [ACC81] [ACO85] [AFL85] [AGOP85]
[AOO82] [AOO86] [ATKI78] [BAKE78] [BROO82] [CARD83a]
[COCK85] [CORM83] [GHEL85] [GRAY78] [HL83] [KNUT73]
[LORI77]

References

[AB85]
   Atkinson, M.P., and Buneman, O.P.,
   "Data Base Programming Language Design."
   Technical Report, University of Glasgow Department of Computer Science, Glasgow,
   Scotland, 1985.

[ABCCM81]
   Atkinson, M.P., Bailey, P.J., Cockshott, W.P., Chisholm, K.J., and Morrison, R.,
   Progress with Persistent Programming,
   Technical Report PPR-8-81, Computer Science Dept., University of Edinburgh.

[ABCCM83]
   Atkinson, M.P., Bailey, P.J., Chisholm, K.J., Cockshott, W.P., and Morrison, R.,
   An approach to persistent programming,
   The Computer Journal, Vol. 26, No. 4, November 1983, pp. 360-365.

[ABCCM84a]
   Atkinson, M.P., Bailey, P.J., Chisholm, K.J., Cockshott, W.P., and Morrison, R.,
   Procedures as Persistent Data Objects,
   Persistent Programming Research Report 9
   Dept. of Computing Science, University of Glasgow & Dept. of Computational
   Science, University of St. Andrews,
   1984.

[ABCCM84b]
   Cockshott, W.P., M.P. Atkinson, K.J Chisholm, P.J. Bailey, and R. Morrison,
   POMS: A persistent Object Management System,
   Software Practice and Experience, Vol. 14, No. 1, January 1984, pp. 49-71.

[ABM85]
   Atkinson, M.P., P. Buneman, and R. Morrison, (Eds.),
   Persistence and Data Types, Papers for the Appin Workshop.
   Persistent Programming Research Report 16,
   University of Glasgow Dept. of Computing Science and
   University of St. Andrews Dept. of Computational Science.

[ABN85]
   Ait-Kaci, H., Boyer, R., and Nasr, R.,
   The Efficient Implementation of Multiple Inheritance.
   MCC Technical Report AI-109-85.
   Microelectronics and Computer Technology Corporation. Austin, December 1985.

266

[ACC81]
Atkinson, M.P., Chisholm, K., and Cockshott, P.,
PS-Algol, an Algol With a Persistent Heap,
Internal Report, CSR-94-81,
University of Edinburgh, Dept. of Computer Science, Edinburgh, 1981.

[ACC83a]
Atkinson, M.P., K.J Chisholm, and W.P. Cockshott,
CMS - A chunk Management System
Software Practice and Experience, Vol. 13, No. 3, March 1983.

[ACC83b]
Atkinson, M.P., K.J Chisholm, and W.P. Cockshott,
Algorithms for a Persistent Heap
Software Practice and Experience, Vol. 13, No. 3, March 1983.

[ACO85]
Albano, A., Cardelli, L., and Orsini, R.,
Galileo: a strongly typed interactive conceptual language
ACM Trans. on Database Systems 10(2), March, 1985

[AF82]
Allen, J.F., and Frish, A.M.,
What's in a Semantic Network,
In Proceedings of the 20th Annual ACL Meeting.
Association for Computational Linguistics, 1982.

[AFL85]
Albano, A., Ferrusi, E.N., and Lucrezia, V.L.,
Un meccanismo per l'interazione con Galileo basato sull'uso di forme,
Atti Congresso AICA 85, Firenze, 1985.

[AGO85]
Albano, A., G. Ghelli, and R. Orsini,
The Implementation of Galileo's Values Persistence,
Proc. of the Persistence and Data Types Workshop, Appin, Scotland, August, 1985.

[AGOP85]
Albano, A., F. Giannotti, R. Orsini, and D. Pedereschi,
The type system of Galileo,
In Proc. 1985 Persistence and Data Types Workshop, Appin, Scotland, August 1985.

[AHU74]
Aho, V.A., Hopcroft, J.E., and Ullman, J.D.,
The Design and Analysis of Computer Algorithms.
Addison-Wesley, 1974.

[AI87]
Arvind, (no initial)., and R.A. Iannucci,
Two fundamental issues in multiprocessing,
In Proceedings of DFVLR - Conference 1987 on Parallel Processing in Science and
Engineering,
Bonn-Bad Godesberg, W. Germany,
June 25-29 1987.

[AITK84a]
Ait-Kaci, H.,
A Lattice-Theoretic Approach to Computation Based on a Calculus of
Partially-Ordered Type Structures.
Ph.D. Thesis, Computer and Information Science,
University of Pennsylvania. Philadelphia, PA,1984.

[AITK84a]
Ait-Kaci, H.,
A Model of Computation Based on Calculus of Type Subsumption.,
PhD thesis, University of Pennsylvania, 1984.

[AITK85]
Ait-Kaci, H.,
Integrating data type inheritance into logic programming.
In Proc. 1985 Persistence and Data Types Workshop, Appin, Scotland, August 1985.

[ALBA83]
Albano, A.,
Type Hierarchies and Semantic Data Models,
ACM SIGPLAN '83: Symposium on Programming Language Issues in Software Systems,
San Francisco, pp. 178-186, 1983.

[ALLC83]
Allchin, J.E.,
An architecture for reliable decentralized systems,
PhD thesis, Georgia Institute of Technology, September 1983,
Available as Technical Report GIT-ICS-83/23.

[ALN86]
Ait-Kaci, H., Lincoln, D., and Nasr, R.,
É : An Overview. MCC Technical Report AI-420-86
(MCC Confidential and Proprietary).
Microelectronics and Computer Technology Corporation. Austin, December 1986.

[ALN87]
Ait-Kaci, H., Lincoln, P., and Nasr R.,
A Logic of Inheritance, Functions, and Equations
MCC Technical Report AI-079-87.
Microelectronics and Computer Technology Corporation.
Austin, Texas. April 1987.
This paper will be presented at the Colloquium on the Resolution of Equations in Algebraic Structures,
organized by Ait-Kaci and Nivat in Lakeway, TX, May 4--6, 1987.

[ALS78]
Anderson, T., Lee, P., and Shrivastava, S.,
A model of recoverability in multilevel systems,
IEEE Transactions on Software Engineering, Vol. SE-4, No. 6, November 1978,
pp. 486-494.

[AM83]
Allchin, J.E., and McKendry, M.S.,
Synchronization and recovery of actions,
In Proceedings of the Second Annual ACM Symposium on Principles of Distributed Computing,
ACM, Montreal, Canada, August 1983, pp. 31-44.

[AM84]
Atkinson, M.P. and Morrison, R.,
Persistent First Class Procedures are Enough,
in Proc. 4th International Conference on the Foundations of Software Technology and Theoretical Computer Science,
Joseph, M., and Shyamasundar, R. (Eds.), Lecture Notes in Computer Science,
No. 181, pp. 223-240, Springer-Verlag, 1984.

[AM85a]
Atkinson, M.P., and Morrison, R.,
Procedures as Persistent Data Objects,
ACM Transactions on Programming Languages and Systems,
7,4, October 1985.

[AM285]
Atkinson, M.P., and R. Morrison,
Types, Bindings, and Parameters in a Persistent Environment,
Proc. of the Persistence and Data Types Workshop, Appin, Scotland, August, 1985.

[AN86a]
Ait-Kaci, H., and Nasr, R.,
Logic Programming and Inheritance.
In Proceedings of the 13th ACM POPL Symposium.
St-Petersburg, Florida, January 1986.

[AN86b]
Ait-Kaci, H., and Nasr, R.,
LOGIN: A Logic Programming Language With Built-In Inheritance.
Journal of Logic Programming 3, pp. 185-215. 1986.

[AN86c]
Ait-Kaci, H., and Nasr, R.,
Residuation: A Paradigm for Integrating Logic and Functional Programming.
MCC Technical Report AI-359-86,
Microelectronics and Computer Technology Corporation. Austin, October 1986.

[AN87a]
Ait-Kaci, H., and Nasr R.,
The Fool's Approach to Computation.
Forthcoming MCC AI/ISA Technical Report,
Microelectronics and Computer Technology Corporation. Austin, 1987.

[AN87b]
Arvind, (no initial)., and R.S. Nikhil,
Executing a program on the mit tagged-token dataflow architecture,
In Proceedings of the PARLE Conference, Eindhoven, The Netherlands.
(LNCS Volume 259), Springer-Verlag, June 15-19 1987.

[ANDE79]
Anderson, T., and Lee, P.,
The provision of recoverable interfaces,
Technical Report No. 137,
University of Newcastle upon Tyne Computing Laboratory,
March 1979.

[ANP87a]
Arvind, (no initial)., R.S. Nikhil, and K.K. Pingali,
Id Nouveau Reference Manual; Part II: Operational Semantics,
Technical Report, MIT Laboratory for Computer Science,
545 Technology Square, Cambridge, MA 02139, USA, April 1987.

[ANP87b]
Arvind, (no initial)., R.S. Nikhil, and K.K. Pingali,
I-Structures: Data Structures for Parallel Computing,
Technical Report Computation Structures Group Memo 269,
MIT Laboratory for Computer Science, 545 Technology Square,
Cambridge, MA 02139, February 1987.
(Also to appear in Proceedings of the Graph Reduction Workshop,
Santa Fe, NM. October 1986.).

[ANS86]
Ait-Kaci, H., Nasr R., and Seo J.,
Babel:A Base for an Experimental Library.
MCC Technical Report AI-379-86.
Microelectronics and Computer Technology Corporation.
Austin, Texas, December 1986.

[AOO82]
Albano, A., Occhiuto,M.E., and Orsini, R.,
A Uniform Management of Persistent and Complex Data in Programming Languages,
In Atkinson, M.P. (editor),
Pergammon Infotech State of The Art Report, Series 9, Number 4: Database, 321-344.
Pergammon Infotech, Maidenhead, England, 1982.

[AOO86]
Albano, A., Occhiuto, M.E., and Orsini, R.,
Galileo Reference Manual, VAX/UNIX Version 1.0,
Servizio Editoriale Universitario di Pisa, January 1986.

[AS81]
Attardi, G., and Simi, M.
Semantics of inheritance and ittributions in the description system Omega
Technical Report A. I. Memo 642, MIT, August, 81.

[ATKI78]
Atkinson, M.P.,
Progamming Languages and Databases,
In S.B.Yao (editor), The Fourth international conference on VLDB,
Berlin, West Germany, 408-419, 1978.

[BA86]
Buneman, P., and Atkinson, M.,
Inheritance and Persistence in Database Programming Languages
In Proc. SIGMOD'86 Conference, pages 4-15, May, 1986

[BAKE78]
Baker, H.G.,
List Processing in Real Time on a Serial Computer,
CACM, 21, 4, 280-294, Apr 78.

[BANC85]
Bancilhon, F.,
Naive evaluation of recursively defined relations.
M.C.C. Technical Report No. DB-004-85, 1985

[BDMN73]
Birtwhistle, G., Dahl, O-J., Myhrhaug, B., and Nygaard, K.,
SIMULA BEGIN, Auerbach Publishers, 1973.

[BFL83]
Brachman, R.J., Fikes, R.E., and Levesque, H.J.,
KRYPTON: A Functional Approach to Knowledge Representation,
FLAIR Technical Report No 16,
Fairchild Lab. for Artificial Intelligence Research,
Fairchild Research Center.
Palo Alto, May 1983.

[BFN82]
Buneman, P., Frankel, R.E., and Nikhil, R.,
An Implementation Technique for Database Query Languages,
ACM Transactions on Database Management 7(2), June, 1982.

[BG80]
Bernstein, P.A., and Goodman, N.,
Timestamp-Based Algorithms for Concurrency Control in Distributed Database System,
Proc. 6th International Conference on Very Large Databases, Montreal, Canada, 1980, pp. 285-300.

[BG83]
Bernstein, P.A., and Goodman, N.,
Multiversion Concurrency Control - Theory and Algorithms,
ACM Transactions on Database Systems, Vol. 8, No. 4, December 1983, pp. 465-483.

[BGL81]
Bernstein, P., Goodman, N., and Lai, M.-Y.,
Two part proof schema for database concurrency control,
In Proceedings of the Fifth Berkeley Workshop on Distributed Data Management and Computer Networks,
February 1981, pp. 71-84.

[BMS80]
Burstall, R.M., D.B. MacQueen, and D.T. Sanella,
Hope: An Experimental Applicative Language,
Technical Report CSR 62-80, Edinburgh Department of Computer Science, 1980.

[BMS84]
Brodie, M.L., J. Mylopoulos, and J.W. Schmidt, (Eds.),
On Conceptual Modelling: Perspectives from Artificial Intelligence, Databases and Programming Languages,
Springer Verlag, Berlin, 1984.

[BMW84]
Borgida, A., J. Mylopoulos, and H.K.T. Wong,
Generalization/specialization as a basis for software specification.
In Brodie, Mylopoulos and Schmidt (Eds.), On Conceptual Modelling,
Springer-Verlag, New York, 1984.

[BORG85]
Borgida, A.,
Language Features for Flexible Handling of Exceptions in Information Systems
ACM Trans. on Database Systems 10(4):565-603, Dec., 1985

[BOUR78]
Bourne, S.R.,
The UNIX Shell,
Bell System Technical Journal, July-August, 1978.

[BR81]
Best, E., and Randell, B.,
A formal model of atomicity in asynchronous systems,
Acta Informatica, Vol. 16, 1981, pp. 93-124.

[BR84]
Brodie, M.L., and D. Ridjanovic,
On the design and specification of database transactions.
In Brodie, Mylopoulos and Schmidt (Eds.), On Conceptual Modelling,
Springer-Verlag, New York, 1984.

[BRIN75]
Brinch Hansen, P.,
The Programming Language Concurrent Pascal,
IEEE Transactions on Software Engineering, Vol. SE-1, No. 2,
June 1975, pp. 199-207.

[BROO82]
Brooker,R.A.,
A 'Database' Subsystem for BCPL,
The Computer Journal, 25, 4, 448-464, 1982.

[BS83]
Bobrow, D.G., and M. Stefik,
The LOOPS Manual,
Xerox PARC, December 1983.

[BSW79]
Bernstein, P.A., Shipman, D.S., and Wong, W.S.,
Formal Aspects of Serializability in Database Concurrency Control,
IEEE Transactions on Software Engineering, Vol. SE-5, No. 3,
May 1979, pp. 203-216

[BUNE84]
Buneman, P.,
Can We Reconcile Programming Languages and Databases?,
In P.M. Stoker, P.M.D. Gray, M.P. Atkinson (Eds.),
Databases - Role and Structure,
Cambridge University Press, Cambridge, UK, 1984.

[BUNE85]
Buneman, P.,
Data types for data-base programming.
In Proc. 1985 Persistence and Data Types Workshop, Appin, Scotland, August 1985.

[CARD83a]
Cardelli,L.,
The Functional Abstract Machine,
AT&T Bell Laboratories Techical Report,TR-107, 1983.
Also in Polimorphism, the ML/LCF/Hope Newsletter , vol 1, no. 1, 1983.

[CARD83b]
Cardelli, L.,
ML Under Unix,
Technical Report, AT&T Bell Laboratories, 1983.

[CARD84]
Cardelli, L.,
A semantics of Multiple Inheritance
In Proc. of Symp. on Semantics of Data Types. Springer Verlag, Sophia Antipolis,
France, June, 1984
(Lecture Notes in CS #173).

[CARD85]
Cardelli, L.,
Amber
In Combinators and Functional Programming Languages, Proc. of the 13th summer
school of LITP. Le Val D'Ajol, Vosges, France, May, 1985.

[CARD86]
Cardelli, L.,
Basic Polymorphism Typechecking, Polymorphism newsletters, II.1, 1986.

[CCA83]
CCA.,
ADAPLEX: Rationale and Reference Manual,
Technical Report CCA-83-03, Computer Corporation of America, 1983.

[CH74]
Campbell, R.H. and Habermann, A.N.
The Specification of process synchronisation by p[ath expressions
Springer-Verlag LNCS Vol 16 1974

[CHAL82]
Challis, M.P.,
Version management - or how to implement transactions without a recovery log,
Database: Infotech State of the Art Report, Vol. 9, No. 8, January 1982,
pp. 435-458.

[CHEN76]
  Chen, P.P.S.,
  The Entity-Relationship Model: Towards a Unified View of Data,
  ACM Transactions on Database Systems, Vol. 1, No. 1, March 1976.

[CHIU82]
  Chiu, G.K.-W.,
  MRDSA User's Manual,
  McGill University, School of Computer Science,
  Technical Report SOCS-82-9, May 1982.

[CHUR40]
  Church, A.,
  A Formulation of the Simple Theory of Types.
  Journal of Symbolic Logic 5,
  pp. 56--68. 1940.

[CM81]
  Clocksin, W.F., and C.S. Mellish,
  Programming in Prolog,
  Springer Verlag, New York.

[CM85]
  Cardelli, L., and D. MacQueen,
  Persistence and Type Abstraction,
  Proc. of the Persistence and Data Types Workshop, Appin, Scotland, August, 1985.

[COCK84]
  Cockshott, W.P.,
  ASSOSARCH - Associator architechture
  Edinburgh University Computing Science Dept, Data Curator Group, 1984.

[COCK85]
  Cockshott, W.P.,
  Addressing Mechanisms and Persistent Programming, (in this volume), 1985.

[CODD70]
  Codd, E.F.,
  A Relational Model for Large Shared Databanks.,
  Communications ACM 13(6):377-387, 1970.

[CODD71a]
  Codd, E.F.,
  Further normalization of the data base relational model.
  In R. Rustin, (Ed.), Data Base Systems,
  Prentice Hall, Engelwood Cliffs, N.J., 1972, pp. 34-64.

[CODD71b]
  Codd, E.F.,
  Relational completeness of data base sublanguages,
  ibid, 65-98.

274

[CODD79]
Codd, E.F.,
Extending the Database Relational Model to Capture More Meaning,
ACM Transactions on Database Systems, Vol. 4, No. 4, December 1979, pp. 397-434.

[CORM83]
Cormack, G.V.,
Extensions to Static Scoping,
ACM SIGPLAN notices, 18, 6, 187-191, Jun 83.

[CW85]
Cardelli, L., and P. Wegner,
On understanding Types, Data Abstraction and Polymorphism,
Computing Surveys, December 1985.

[DATE83a]
Date, C.J.,
An Introduction to Database Systems Volume II,
Addison-Wesley, Reading, Mass., 1983.

[DATE83b]
Date, C.J.,
The outer join,
In Proc. 1983 2nd International Conference on Databases (ICOD-2),
S.M. Deen and P. Hammersley, (Eds.),
Cambridge, U.K., September 1983, pp. 76-106.

[DATE86]
Date, C.J.,
An Introduction to Database Systems, Volume I (Fourth Edition),
Addison Wesley, Reading, MA, USA, 1986.

[DAVI73]
Davies, C. T.,
Recovery Semantics for a DB/DC System,
In Proceedings of the ACM Annual Conference, ACM, Atlanta, GA, 1973, pp. 136-141.

[DAVI78]
Davies, C.T.,
Data processing spheres of control,
IBM System Journal, Vol. 17, No. 2, 1978, pp. 179-198.

[DD79]
Demers, A., and Donahue, J.,
Revised Report on Russell,
TR 79-389, Dept. of Computer Science, Cornell University.

[DD80a]
Demers, A., and Donahue, J.,
The Semantics of Russell: An Excercise in Abstract Data Types.,
Technical Report, Computer Science Department, Cornell University, 1980.

[DD80b]
  Demers, A.J., and J.E. Donahue,
  Data Types, Parameters and Type Checking,
  In Proceedings 1980 of the 7th Symposium on the Principles of Programming
  Languages,
  ACM.

[DH72]
  Dahl, D., and C. A. R. Hoare,
  Hierarchical Program Structures,
  In Dahl, Dijkstra, Hoare, Structured Programming, Academic Press, 1972.

[DHIL85]
  Dhillon, R.,
  Modifications to MRDS/FS,
  McGill University.

[DIJK68]
  Dijkstra, E.W.,
  Cooperating Sequential Processes,
  in Programming Languages, Genuys, (no initial) (Ed.), Academic Press, 1968.

[DK79]
  Deliyanni, A., and Kowalski, R.A.
  Logic and Semantic Networks,
  Communications of the ACM,
  22(3):184-92. 1979.

[DN66]
  Dahl, O., and Nygaard, K.,
  Simula, an Algol-based Simulation Language,
  Communications on the ACM, Vol. 9, pp. 671-678, 1966.

[DUEC83]
  Duechting, B.,
  A relational picture editor,
  McGill University, School of Computer Science
  Technical Report SOCS-83-19, August 1983.

[EEKMS85]
  Eckhardt, H., Edelmann, J., Koch, J., Mall, M. and Schmidt, J.W.
  Draft Report on the database programming language DBPL
  Johann Wolfgang Goethe University,
  Frankfurt on Maine West Germany. 1985.

[EGLT76]
  Eswaran, K.P., Gray, J.N., Lorie, R.A., and Traiger, I.L.,
  The notions of consistency and predicate locks in a database system,
  Communications of the ACM, Vol. 19, No. 11, November, 1976, pp. 624-633.

[FAIR82]
  Fairbairn, J.,
  Ponder and its type system,
  Cambridge University Computer Laboratory Technical Report 31, 1982.

276

[FAIR85]
   Fairbairn, J.,
   A New Type-Checker for a Functional Language,
   In Proc. 1985 Persistence and Data Types Workshop, Appin, Scotland, August 1985.

[FI73]
   Falkoff, A.D., and Iverson, K.E.,
   The design of APL,
   IBM Journal of Research and Development, Vol. 10, No. ?, July 1973, pp. 324-334.

[FIND79]
   Findler, N.V., (Ed.),
   Associative Networks: Representation and Use of Knowledge by Computers.
   Academic Press. New York, NY. 1979.

[GHEL85]
   Ghelli, G.,
   La gestione della persistenza dei valori nel linguaggio Galileo,
   Tesi di laurea in Scienze dell'Informazione, Pisa, 1985.

[GM78]
   Gallaire, H., and Minker, J. (Eds.),
   Logic and Databases,
   Plenum Press,
   New York, 1978.

[GMN84]
   Gallaire, H., J. Minker, and J-M. Nicolas,
   Logic and databases: a deductive approach,
   ACM Computing Surveys, Vol. 16, No. 2, June 1984, pp. 153-185.

[GMW79]
   Gordon, M.J., Milner, A.J.R.G., and Wadsworth, C.P.,
   Edinburgh LCF,
   Lecture Notes in Computer Science, No. 78,
   Springer-Verlag, New York, 1979.

[GOGU83]
   Goguen, J.,
   Report on ADA Program Libraries Workshop,
   SRI International, 1983.

[GR83]
   Goldberg, A., and Robson, D.,
   Smalltalk-80: The Language and its Implementation,
   Addison-Wesley, 1983.

[GRAY78]
   Gray, J.N.,
   Notes on Database Operating Systems,
   In Operating Systems: an Advanced Course, Bayer, R., et al. Eds.,
   Springer-Verlag, Berlin, 393-481, 1978.

[HARP85]
  Harper, R.,
  Modules and Persistence in Standard ML,
  In Proc. 1985 Persistence and Data Types Workshop, Appin, Scotland, August 1985.

[HARP86]
  Harper, R.,
  Introduction to Standard ML,
  Computer Science Department, University of Edinburgh, September 1986.

[HB80]
  Hammer, M., and Berkowitz, B.,
  Dial: A programming language for data-intensive applications
  In Proc. ACM SIGMOD Conference, pages 140-173, May, 1980

[HL83]
  Hewitt, C., and Lieberman, H.,
  A Real-Time Garbage Collector Based on the Lifetime of Objects,
  CACM, 26, 6, 419-429, Jun 83.

[HM81]
  Hammer, M., and McLeod, D.,
  Database Description with SDM: A Semantic Database Model,
  ACM Transactions on Database Systems, Vol. 6, No. 3, September 1981.

[HMM86]
  Harper, R., D. MacQueen, and R.Milner,
  Standard ML,
  Computer Science Department, University of Edinburgh, March 1986.

[HOAR74]
  Hoare, C.A.R.,
  Monitors: an operating system structuring Concept,
  Connumications of the ACM, Vol. 17, No. 10, October 1974, pp. 549-557.

[HUET76]
  Huet, G.,
  Resolution d'Equations dans des Langages d'Ordre 1, 2, ..., W.
  These de Doctorat d'Etat,
  Universite de Paris VII, France. 1976.

[IBM]
  System 38.,
  journal-article

[ICHB83]
  Ichbiah et al,
  The Programming Language ADA Reference Manual
  ANSI/MIL-STD-1815A-1983 (1983)

[JW78]
  Jensen, P. and Wirth, N.
  Pascal User Manual and Report,
  Springer Verlag LNCS 1978

[KAEH81]
  Kaehler, T.,
  Virtual Memory for an Object Oriented Language,
  Byte, August 1981.

[KMPRSZ83]
  Koch, J., Mall, M., Putfarken, P., Reimer, M., Schmidt, J.W. and Zehnder, C.A.
  Modula/R Report
  Technical Report, Eidgenossische Technische Hochschule Zurich,
  Institute Fur Informatik. 1983

[KNUT73]
  Knuth, D.E.,
  The Art of Computer Programming, Vol.1: Fundamental Algorithms,
  Addison-Wesley, Reading, Mass., 1973.

[KNUT79]
  Knuth, D.E.,
  Tex and Metafont, New Directions in Typesetting,
  Digital Press/American Mathematical Society, 1979.

[KORT81]
  Korth, H.F.,
  Locking protocols: general lock classes and deadlock freedom,
  PhD thesis, Princeton University, 1981.

[KOWA78]
  Kowalski, R.A.,
  Logic for Data Description,
  In Logic and Data Bases,
  Gallaire, H., and Minker, J. (Eds.),
  pp.77-103. Plenum Press, 1978.

[KR81]
  Kung, H.T., and Robinson, J.T.,
  On Optimistic Methods for Concurrency Control,
  ACM TODS# 6(2), pp 213-226, June 1981.
  ACM Transactions on Database Systems, Vol. 6, No. 2, June 1981, pp. 213-226.

[KRAB75]
  Krablin, G.L.,
  Experimental Concurrency in PS-algol,
  unpublished report, Dept. of Computing Science, University of Glasgow, 1985.

[LAMP81]
  Lampson, B.W.,
  Atomic Transactions,
  in Distributed Systems - Architecture and Implementation (LNCS 105), ed. Lampson,
  Paul, and Seigert, pp. 246-265, Springer-Verlag, 1981.

[LAND66]
  Landin, P.J.
  The Next 700 Programming Languages
  CACM 9,3 pp 157-164

[LHJLSW83]
Liskov, B., Herlihy, M., Johnston, P., Leavens, G., Scheifler, R. and Weihl, W.
Preliminary Argus Manual
Technical Report Memo 39 (1983) M.I.T.

[LISK81]
Liskov, B., et al.,
CLU reference manual,
In Goos, and Hartmanis, (Eds.), Lecture Notes in Computer Science, Vol. 114,
Springer-Verlag, Berlin, 1981.

[LISK82]
Liskov, B.,
On linguistic support for distributed programs,
IEEE Transactions on Software Engineering, Vol. SE-6, No. 3, May 1982, pp. 203-210.

[LS83]
Liskov, B., and Scheifler, R.,
Guardians and actions: linguistic support for robust, distributed programs,
ACM Transactions on Programming Languages and Systems, Vol. 5, No. 3, July 1983,
pp. 381-404.

[LISK84]
Liskov, B.,
Overview of the ARGUS Language and System,
MIT Programming Methodology Group Memo 40, February 1984.

[LORI77]
Lorie, A.L.,
Physical Integrity in a Large Segmented Database,
ACM TODS, 2, 1, 91-104, Mar 77.

[LNCS84]
International Symposium on the Semantics of Data Types., Sophia-Antipolis,
France (Springer-Verlag LNCS 173), June 1984.

[LYNC82]
Lynch, N.A.,
Concurrency control for resilient nested transactions,
In Proceedings of the Second ACM Symposium on Principles of Database Systems,
March 1983.

[LZ74]
Liskov, B., and Zilles, S.N.,
Programming with abstract data types,
In Sigplan Notices,
Proceedings of the ACM SIGPLAN Conference on Very High Level Languages,
ACM, Vol. 9, No. 4, April 1974, pp. 50-59.

[MACQ85]
MacQueen, D.B.,
Modules for Standard ML,
In Polymorphism II.2, October 1985.

280

[MACQ86]
MacQueen, D.B.,
Using dependent types to express modular structure,
In Proc. 1986 13th ACM Symposium on Principles of Programming Languages,
St. Petersburg Beach, Florida, January 1986, pp. 277-286.

[MATT85a]
Matthews, D.C.J.,
Poly Manual, SIGPLAN Notices, Vol. 20, No. 9, September 1985.

[MATT85]
Matthews, D.C.M.,
An Overview of the Poly Programming Language,
Proc. of the Persistence and Data Types Workshop, Appin, Scotland, August, 1985.

[MBW80]
Mylopoulos, J., P.A. Bernstein, and H.K.T. Wong,
A Language Facility for Designing Database-Intensive Applications,
ACM Transactions on Database Systems, Vol. 5, No. 2, June 1980, pp. 185-207.

[MCAR62]
McCarthy et al.
Lisp 1.5 Programmers Manual
M.I.T. Press Cambridge Massechussetts

[MD85]
Merrett, T.H.,
Relational storage and processing of two-dimensional diagrams,
Computers and Graphics, Vol. 9, No. 3.

[MERR77]
Merrett, T.H.,
Relations as programming language elements,
Info. Proc. Lett., Vol. 6, No. 1, February 1977, pp. 29-33.

[MERR84]
Merrett, T.H.,
Relational Information Systems,
Reston Publishing Co., Reston, Va., 1984.

[MERR85a]
Merrett, T.H.,
First steps to algebraic processing of text,
In G. Gardarin and E. Gelenbe (Eds.),
New Applications of Data Bases, Academic Press, 1985.

[MERR85]
Merrett, T.H.,
Persistence and aldat,
In Proc. 1985 Persistence and Data Types Workshop, Appin, Scotland, August 1985.

[MILN78]
Milner, R.,
A Theory of Type Polymorphism in Programming,
Journal of Computer and System Sciences, Vol. 17, No. 3, December 1978.

[MILN80]
Milner, R.
A Calculus of Communicating Systems
Springer-Verlag LNCS 8,1 1980

[MILN84]
Milner, R.,
A proposal for Standard ML,
In Proc. of the 1984 ACM Symposium on Lisp and Functional Programming,
Austin, Texas, August 1984.

[MILN85]
Milner, R.,
The Standard ML Core Language, in [HARP86a].

[MINS75]
Minsky, M.,
A Framework for Representing Knowledge.
In Winston, P.H. (Ed.),
The Psychology of Computer Vision,
pp. 211--77. McGraw-Hill. New York, NY. 1975.

[MITC79]
Mitchell, J.G. et al.,
MESA Language Manual, XEROX PARC, 1979

[MM79]
McSkimin, J.R., and Minker, J.,
A Predicate Calculus Based Semantic Network for Question-Answering Systems,
In Associative Networks---The Representation and Use of Knowledge by Computers,
Findler, N. (Ed.).,
Academic Press, New York, 1979.

[MOSS81]
Moss, J.E.B.,
Nested transactions: an approach to reliable distributed computing,
PhD thesis, MIT, 1981,
Available as Technical Report MIT/LCS/TR-260.

[MP84]
Meertens, L.G.L.T., and S. Pemberton,
An implementation of the B programing language,
Centrum voor Wiskunde en Informatica, 1984.

[MP85]
Mitchell, J.C., and G.D. Plotkin,
Abstract Types have Existential Type,
In Proc. 1985 12th ACM Symposium on Principles of Programming Languages,
New Orleans, Louisiana, January 1985, pp. 37-51.

[MS82]
MacQueen, D.B., and R. Sethi,
A Semantic Model of Types for Applicative Languages,
Symposium on Lisp and Functional Programming, ACM 1982.

282

[MSP84]
MacQueen, D.B., R. Sethi, and G. Plotkin,
An Ideal Model for Recursive Polymorphic Types,
Eleventh Annual ACM Symposium on Principles of Programming Languages, 1984.

[MW80]
Mylopolous, J. and Wong, H.K.T.
Some Features of the TAXIS Data Model
6th International Conference on Very Large Data Bases (1980) Montreal.

[NAUR63]
Naur, P.
Revised Report on the Algorithmic Language Algol 60
CACM 6,1 pp 1-17

[NIKH84]
Nikhil, R.S.,
An Incremental, Strongly-Typed Database Query Language,
PhD thesis, Moore School, University of Pennsylvania,
Philadelphia, PA, August 1984.
Available as Technical Report MS-CIS-85-02.

[NIKH87]
Nikhil, R.S.,
Id Nouveau Reference Manual, Part I: Syntax,
Technical Report, Computation Structures Group, MIT Lab. for Computer Science,
545 Technology Square, Cambridge, MA 02139, April 1987.

[ORGA84]
Organick, E.,
A programmers view of the Intel 432,
McGrawHill, 1984.

[PS85]
PS-algol Reference Manual., Fourth Edition,
Persistent Programming Research Report 12,
Dept. of Computing Science, University of Glasgow and Dept. of Computational
Science, University of St. Andrews,
1985.

[REED78]
Reed, D.P.,
Naming and synchronization in a decentralized computer system,
PhD thesis, MIT, 1978,
Available as Technical Report MIT/LCS/TR-205.

[REIS84]
Reiss, S.,
Graphical Program Development with PECAN Program Development System,
Brown University, Department of Computer Science, Technical Report No. CS-84-04.

[REISS]
Reiss, S.,
forthcoming Brown University, Department of Computer Science, Technical Report on
the GARDEN programming environment.

[REIT84]
  Reiter, R.,
  Towards a logical reconstruction of relational database theory.
  In Brodie, Mylopoulos and Schmidt (Eds.), On Conceptual Modelling,
  Springer-Verlag, New York, 1984.

[REYN70]
  Reynolds, J.C.,
  Transformational Systems and the Algebraic Structure of Atomic Formulas,
  In Machine Intelligence 5,
  Michie, D. (Ed.),
  Edinburgh University Press, 1970.

[REYN85]
  Reynolds, J.C.,
  Three Approaches to Type Structure,
  TAPSOFT Advanced Seminar on the Role of Semantics in Software Development,
  Berlin, March 25-29, 1985.

[REYN86]
  Reynolds, J.C.,
  Three Approaches to Type-Structure,
  In Springer Lecture Notes in Computer Science No. 185, 1986.

[ROBI65]
  Robinson, J.A.,
  A Machine Orientated Logic Based on the Resolution Principle,
  Journal of the ACM, Vol. 12, 1965, pp 23-31.

[ROWE80]
  Rowe, L.
  Reference Manual for the Programming Language RIGEL
  Technical Report of the University of California at Berkeley.

[RS79]
  Rowe, L., and Shoens, K.,
  Data Abstraction, Views and Updates in RIGEL,
  Proc. 1979 ACM SIGMOD International Conference on Management of Data,
  Boston, Mass., May 1979, pp. 71-81.

[SCHM77]
  Schmidt, J.W.,
  Some high-level language constructs for data of type relation,
  ACM Transactions on Database Systems, Vol. 2, No. 3,
  September 1977, pp. 247-261.

[SCHM78]
  Schmidt, J.W.
  Type Concepts for Database Definition
  in Datbases: Improving their usability and responsiveness, Academic Press (1978).

[SCHW84]
  Schwarz, P.,
  Transactions on typed objects,
  PhD thesis, CMU, December 1984,
  Available as Technical Report CMU-CS-84-166.

[SCOT82]
Scott, D.,
Domains for Denotational Semantics.,
In ICALP 1982, Aarhus, Denmark. July, 1982.

[SF]
Saib, S.H., and R.E. Fritz,
The ADA Programming Language: A Tutorial,
IEEE Computer Society Press, IEEE Catalog Number EHO 202-2.

[SFL81]
Smith, J.M., Fox, S., and Landers, T.,
Reference Manual for ADAPLEX,
Technical Report CCA-81-02,
Computer Corporation of America, Cambridge, Mass., January 1981.

[SFL83]
Smith, J.M., Fox, S., and Landers, T.,
ADAPLEX: Rationale and Reference Manual second edition,
Computer Corporation of America,
Four Cambridge Center, Cambridge, Massachusetts 02142, 1983.

[SHIP81]
Shipman, D.W.,
The Functional Data Model and the Data Language DAPLEX.,
ACM Transactions on Database Systems 6(1):140-173, March, 1981.

[SHOP79]
Shopiro, J.E.
THESEUS - a programming language for relational databases
ACM TODS 4,4 (1979)

[SOLT81]
Soltis, F.G.,
Design of a Small Business Data Processing System,
IEEE Computer, September 1981.

[SS77]
Smith, J.M., and Smith, D.C.P.,
Database Abstractions - Aggregation and Generalisation,
ACM Transactions on Database Systems, Vol. 2, No. 2, June 1977, pp. 105-133.

[SS84]
Schwarz, P., and Spector, A.,
Synchronizing shared abstract types,
ACM Transactions on Computer Systems, Vol. 2, No. 3, August 1984.

[STON84]
Stonebraker, M.,
A database perspective,
In Brodie, Mylopoulos and Schmidt (Eds.), On Conceptual Modelling,
Springer-Verlag, New York, 1984.

[TENN77]
Tennent, R.D.
Language Design Methods Based on Semantic Principles
Acta Informatica 8 (1977) pp 97-112

[TURN79]
Turner, D.QA.
SASL Language Manual,
University of St. Andrews CS/79/3 (1979)

[TURN81]
Turner, D.A.,
The semantic elegance of applicative languages,
In Proc. ACM Conference on Functional Programming Languages and Computer
Architecture,
Portsmouth, New Hampshire, pages 85-92, ACM, October 1981.

[ULLM82]
Ullman, J.D.,
Principles of Database Systems. ,
Pittman, 1982. Second Edition.

[VANR83]
Van Rossum, T.,
*Implementation of a domain algebra and a functional syntax for
a relational database system,*
McGill University, School of Computer Science,
Technical Report SOCS-83-18, August 1983.

[VANW75]
van Wijngaarden, A., et al,
The Revised Report on the Algorithmic Language Algol 68,
Springer Verlag 1975

[VERH76]
Verhofstad, J.S.M.,
Recovery for multi-level data structures,
Technical report 96, University of Newcastle upon Tyne, December 1976.

[WARR83]
Warren, D.H.D.,
An Abstract Prolog Instruction Set,
AI Center Technical Note 309,
SRI International, Menlo Park, CA., October 1983.

[WEIH83]
Weihl, W.E.,
Data-dependent concurrency control and recovery,
In Proceedings of the Second Annual ACM Symposium on Principles of Distributed
Computing,
ACM, Montreal, Canada, August 1983, pp. 63-75.

[WEIH84]
Weihl, W.E.,
Specification and implementation of atomic data types,
PhD thesis, MIT, 1984,
Available as Technical Report MIT/LCS/TR-314.

286

[WEIH85]
Weihl, W.E.,
Linguistic Support for Atomic Data Types,
in Proceedings Appin Workshop on Persistence and Data Types, Appin, Scotland,
August 1985.

[WIRT71]
Wirth, N.
The Programming Language Pascal
Acta Informatica 1,1 (1971) pp 35-63

[WL85]
Weihl, W., and Liskov, B.,
Implementation of resilient, atomic data types,
ACM Transactions on Programming Languages and Systems, April 1985.

[WSKRV81]
Wasserman, A.I., Shertz, D.D., Kersten, M.L., Reit, R.P., and van de Dippe, M.D.,
Revised Report on the Programming Language PLAIN,
ACM SIGPLAN Notices, 1981

[XERO81]
The Xerox Learning Research Group.,
The Smalltalk-80 system.,
Byte 6:36-48, August, 1981.

[ZDON84]
Zdonik, S.B.,
Object Mangement System Concepts,
Proceedings of the Second ACM-SIGOA Conference on Office Information Systems,
Toronto, Canada, June, 1984.

[ZW85]
Zdonik, S.B., and P. Wegner,
A database approach to languages, libraries and environments,
In Proc. 1985 Persistence and Data Types Workshop, Appin, Scotland, August 1985.

[ZW85]
Zdonik, S.B., and P. Wegner,
A Database Approach to Languages, Libraries, and Environments,
Brown University, Department of Computer Science,

INDEX

List of Authors

Aït-Kaci, H.
Microelectronics and Computer
Technology Corporation,
9430 Research Boulevard,
Austin, TX78759,
USA.

Albano, A.
University of Pisa,
Instituto di Science del Informazione,
Corso, Italia 40,
I-56100,
Pisa, Italy.

Atkinson, M.P.
Department of Computing Science,
University of Glasgow,
14 Lilybank Gardens,
Glasgow, G12 8QQ,
Scotland.

Borgida, A.
Department of Computer Science,
Rutgers University,
New Brunswick, New Jersey,
USA.

Buneman, P.
Department of Computer Information
Science, The Moore School of Electrical
Engineering,
University of Pennsylvania,
Philadelphia 19104-3897,
USA.

Cardelli, L.
AT&T Bell Laboratories,
Room 2C 322,
Murray Hill, NJ07974,
USA.

Cockshott, W.P.
MEMEX,
21 Landsdowne Crescent,
Edinburgh, Scotland.

Fairbairn, J.
Computer Laboratory,
University of Cambridge,
Corn Exchange Street,
Cambridge, CB2 3QG,
England.

Ghelli, G.
University of Pisa,
Instituto di Science del Informazione,
Corso, Italia 40,
I-56100,
Pisa, Italy.

Giannotti, F.
University of Pisa,
Instituto di Science del Informazione,
Corso, Italia 40,
I-56100,
Pisa, Italy.

Harper, R.
Department of Computer Science,
University of Edinburgh,
Edinburgh, EH9 3JZ,
Scotland.

Krablin, G.L.
Burroughs Corporation,
PO Box 203,
Paoli, PA 19310,
USA.

MacQueen, D.
AT&T Bell Laboratories,
Room 2C 322,
Murray Hill, NJ07974,
USA.

Matthews, D.C.J.
Computer Laboratory,
University of Cambridge,
Corn Exchange Street,
Cambridge, CB2 3QG,
England.

Merrett, T.
School of Computer Science,
McGill University,
805 Sherbrooke Street West,
Montreal, Quebec, H3A 2K6,
Canada.

Morrison, R.
University of St Andrews,
Department of Computational Science,
North Haugh, St Andrews,
Fife, KY16 9SS,
Scotland.

Nikhil, R.S.
33 Fayette Street,
Boston, Massachusetts,
USA.

Orsini, R.
University of Pisa,
Instituto di Science del Informazione,
Corso, Italia 40,
I-56100,
Pisa, Italy.

Pedreschi, D.
University of Pisa,
Instituto di Science del Informazione,
Corso, Italia 40,
I-56100,
Pisa, Italy.

Wegner, P.
Brown University,
Department of Computer Science,
Box 1910,
Providence, RI 02912,
USA.

Weihl, W.
Laboratory for Computer Science,
Massachusetts, Institute of Technology,
545 Technology Square,
Cambridge, Massachusetts,
USA.

Zdonik, S.B.
Brown University,
Department of Computer Science,
Box 1910,
Providence, RI 02912,
USA.

Springer-Verlag
Berlin Heidelberg New York
London Paris Tokyo

Springer

Topics in Information Systems

Series Editors:
M. L. Brodie, J. Mylopoulos,
J. W. Schmidt

D. Tsichritzis

Office Automation

Concept and Tools

1985. 86 figures. XII, 441 pages. ISBN 3-540-15129-X

Contents: Integration. – Filing. – Mailing. – Procedure Specification. – Modelling. – Analysis. – Performance. – Epilogue. – References. – Index.

W. Kim, D. S. Reiner, D. S. Batory (Eds.)

Query Processing in Database Systems

1985. 127 figures. XIV, 365 pages. ISBN 3-540-13831-5

Contents: Introduction to Query Processing. – Query Processing in Distributed Database Management Systems. – Query Processing for Multiple Data Models – Database Updates through Views. – Database Access for Special Applications. – Techniques for Optimizing the Processing of Multiple Queries. – Query Processing in Database Machines. – Physical Database Design. – References. – List of Authors. – Subject Index.

M. L. Brodie, J. Mylopoulos, J. W. Schmidt (Eds.)

On Conceptual Modelling:

Perspectives from Artificial Intelligence, Databases, and Programming Languages

1984. 25 figures. XI, 510 pages. ISBN 3-540-90842-0

Contents: Artificial Intelligence, Database, and Programming Language Overviews. – Perspectives from Artificial Intelligence. – Perspectives from Databases. – Perspectives from Programming Languages. – Concluding Remarks from Three Perspectives. – References. – Authors and Symposium Participants. – Index.

Springer-Verlag
Berlin Heidelberg New York
London Paris Tokyo